T0248655

*Freud on the Acropolis*

# Freud on the Acropolis

## Reflections on a Paradoxical Response to the Real

*Susan Sugarman*

Routledge
Taylor & Francis Group

LONDON AND NEW YORK

First published 1998 by Westview Press

Published 2018 by Routledge
52 Vanderbilt Avenue, New York, NY 10017
2 Park Square, Milton Park, Abingdon, Oxon OX14 4RN

*Routledge is an imprint of the Taylor & Francis Group, an informa business*

Copyright © 1998 Taylor & Francis

All rights reserved. No part of this book may be reprinted or reproduced or utilised in any form or by any electronic, mechanical, or other means, now known or hereafter invented, including photocopying and recording, or in any information storage or retrieval system, without permission in writing from the publishers.

Notice:
Product or corporate names may be trademarks or registered trademarks, and are used only for identification and explanation without intent to infringe.

Library of Congress Cataloging-in-Publication Data
Sugarman, Susan.
   Freud on the Acropolis : reflections on a paradoxical response to
the real / Susan Sugarman.
      p.   cm.
   Includes bibliographical references and index.
   ISBN 0-8133-9088-5 (hardcover)
   1. Psychoanalysis.   2. Freud, Sigmund, 1856–1939.   I. Title.
BF173.S847   1998
150.19'5—dc21                                                                98-11324
                                                                                    CIP

ISBN 13: 978-0-367-01509-1 (hbk)
ISBN 13: 978-0-367-16496-6 (pbk)

# *Contents*

*Acknowledgments*      vii

Introduction      1

1   The Problem      5

2   A Synopsis of Freud's Account      13

3   Freud's Thought at Face Value      17

    1. Freud's Thought as a Literal Expression, 18
    2. Literal Interpretations of Other Thoughts, 23

4   Freud's Thought as a Figurative Expression      25

    1. Linguistic Tropes, 25
    2. Indirect Discourse, 26
    3. Loose Usage, 26
    4. Summary of Common-Sense Interpretation
       (Chapters 3 and 4), 27

5   Freud's Account Analyzed      29

    1. Freud's Account of His Experience, 30
    2. The General Case of the Thought, 32

6   A Different Neurosis      35

    1. A Preliminary Account, 35
    2. What If . . . ? 36
    3. A Pure Affirmation, 37

7   The Thought as Mundane      39

    1. Miracles and Lesser Events, 39
    2. Further Examples of the Thought, 41

8 The Thought as Childlike: The Reencounter
with the Known in a New Guise ......... 47

  1. The Childlike Vantage Point of the Thought, 48
  2. Adult Contexts for the Return of the Childlike, 50

9 The Reencounter with the Known in Reality ......... 57

  1. Connecting with the Object's Connections, 58
  2. Connecting with the Personal Past, 59
  3. Freud's Thought Revisited, 62

10 The Adult Voice ......... 65

  1. Children and the Thought, 65
  2. The Confirmation of Experience, 68

11 The Paradoxical Stance ......... 73

*Appendix: "A Disturbance of Memory on the Acropolis,"*
  Sigmund Freud (translated by James Strachey) ......... 79
*Notes* ......... 87
*References* ......... 99
*Index* ......... 105

# Acknowledgments

I gratefully acknowledge a fellowship I received in 1989–1990 from the John Simon Guggenheim Memorial Foundation and an Old Dominion Faculty Fellowship from the Humanities Council at Princeton University from 1996 to 1998 for their support of portions of my work on this book.

I thank the following people for real or hypothetical examples that appear in the book or that, although not included, influenced arguments that do appear: Rüdiger Bittner, Harry Frankfurt, Sam Glucksberg, Peter Gollwitzer, Richard Gonzales, Charles Gross, Saul Kripke, Alan Leslie, Perry Link, Michael Maratsos, George Miller, Toni Pickard, Eric Santner, and Irene Vogel. For additional source material I thank Richard Beckwith, Greta Berman, Jerome Bruner, Sam Glucksberg, Charles Gross, Venā Jordan, Zena Meadowsong, James Richardson, Donald Spence, and Ann Himmelberger Wald.

For comments on the manuscript or other general discussion, I thank Rüdiger Bittner, Jerome Bruner, Margaret Carr, Jonathan Cobb, Joel Cooper, Stanton de Riel, Catherine Elgin, Michael Frede, Ellyn Geller, David Getson, Sam Glucksberg, Gilbert Harman, Philip Johnson-Laird, Harold Langsam, Jonathan Lear, Richard Moran, James Richardson, Jacqueline Savani, Robert Schwartz, and Ann Himmelberger Wald. For discussion I am grateful also to meetings, respectively, of the Cognitive Psychology Research Seminar and the Old Dominion Faculty Fellows, both at Princeton. I thank Catherine Pusateri and Marcus Boggs of Westview Press for their sensitivity and insight in helping me to work the manuscript into its final form.

To all of these people and the many others who have sat and discussed "Freud's thought" with me, I owe thanks not only for their contribution to the book, but also for the inspiration and encouragement they provided as the book developed. I also thank the students at Princeton who, while attending my seminars on Freud over the last several years, continually pushed me—and each other—to probe, and to find the joy in doing so.

*Susan Sugarman*
Princeton University

*Also existiert das alles wirklich so wie wir es*
*auf der Schule gelernt haben?!*
S. Freud, Brief an Romain Rolland:
Eine Erinnerungesstörung auf der Akropolis, 1936

*So all this really does exist,*
*just as we learned at school!*
S. Freud, A disturbance of memory on the Acropolis:
An open letter to Romain Rolland
on the occasion of his seventieth birthday, 1936
(J. Strachey, trans.)

# Introduction

Some have described the study of mind in the 1990s as a turf war between neuroscience and philosophy. Do significant questions remain to be asked about the nature of mental life, the combatants wonder, apart from the question of where different mental functions are localized in the brain? Is there any activity to be scrutinized, apart from the "activity" of the neural substrates that may govern these functions?[1] The controversy had a parallel in the 1980s in debates over whether various computer models meaningfully or appropriately simulate human thought and feeling.[2] Before that, people disputed whether human mental life reduces essentially to stimulus-response connections, or whether one must stipulate the intervention of an active cognitive "processor." If the latter, the contestants asked, then how would one model the processor?[3]

The search for the appropriate "model" of the mind, or alternatively for its neural substrates, presupposes that we know *what* aspects of mind should be modeled or for which ones neural substrates should be sought. Even to conclude that the nature of mind is *not* well illuminated by given models presupposes that human thought, feeling, and experience are largely known and remain mostly to be characterized optimally.

We know many things about our mental life, but we do not know all. Room still exists for the enduring psychological questions: Why do we think, feel, see, and do, as we do? *What* do we think, feel, see, and do—in human terms?

Sigmund Freud devoted the greater part of his career to the pursuit of these questions. In doing so, he began his investigations with the description of potentially informative phenomena:

> The view is often defended that sciences should be built up on clear and sharply defined basal concepts. In actual fact no science, not even the most exact, begins with such definitions. The true beginning of scientific activity consists rather in describing phenomena and then in proceeding to group, classify, and correlate them.[4]

Freud is often associated with the opposite approach. Popularizations characterize him as an inquirer committed to the defense of a theory and blinded by it in his view of phenomena. But the conception of science that he describes in the preceding passage reflects accurately a method that he followed at least in outline in his writings. Many of his inquiries begin with the identification of some particular phenomenon—a behavior, thought, or feeling—that poses a puzzle of some kind. Freud set out to solve the puzzle.

Thus he centered inquiries around apparently normal, but to his mind puzzling, phenomena such as the experience of the uncanny, slips of the tongue, jokes, and people's loss of intelligence and moral integrity in groups, as well as around patently abnormal phenomena such as neurotic symptoms.[5] Why, Freud asked for example, do people show diminished intellectual and moral sensibility in groups (the extreme manifestation of this tendency occurring in mobs)? From a naïve perspective, one might think that people's mental functioning might be *enhanced* by the participation in a collective.[6] Alternatively, what is the content of the feeling of uncanniness, the feeling that arises when, for instance, the same coincidence repeatedly occurs or an inert object seems to come alive?[7] Although most people recognize the feeling, it is difficult to describe.

To solve these puzzles, Freud, at his best, proceeded as would any scientist who was trying to construct a theory to fit data. He formulated and tested hypotheses. He conducted the tests by projecting the likely outcome of the hypotheses, predicting the form that people's behavior ought to take if the hypothesis were true, and then observing whether the outcome obtained.

Freud reasons, for example, that if the feeling of uncanniness reduced to a feeling of dread, as common intuition might suggest that it does, then human languages ought not to distinguish the uncanny from other feelings of dread. But they do. Similarly, if the feeling originated in intellectual uncertainty as to whether something is dead or alive, as some writers of his day suggested that it did, then all occurrences of uncanny feeling ought to include this uncertainty. Some occurrences, however, do not appear to do so. For instance, people find it uncanny when someone of whom they have just been thinking suddenly appears, or when the same, seemingly accidental event repeatedly befalls them. One returns several times to the same spot when one is lost, for example, or recurrently encounters the same number on a guest check, in a bakery queue, on a theater ticket, and so on.

Freud determines, after further analysis, that the feeling of uncanniness is a special case of dread. It results from the revival by a contemporary event of a wish, feeling, or other impression dating from one's earlier development that has since been either repressed (ejected from consciousness because of its painful content) or outgrown. The wish or other impression frequently assumes an animistic premise, a premise that fuses psychological and physical causation or that exaggerates the power of human thought and action, for example. The quality of uncanniness arises from the confluence of archaic impressions of this sort, now revived, on the one hand, and one's sense that these impressions have been superseded, on the other hand.

Freud was efficient and shrewd in his application of critical analysis to the task of eliminating unworkable hypotheses. Compelling stretches of argumentation of this kind appear, for example, in his books on jokes and on the "psychopathology of everyday life" (in which he treats slips of the tongue and other small errors of thought, speech, and action), as well as in some of his discussions of particular neurotic symptoms.[8]

Some might fault Freud for not having carried the critical process far enough, for abandoning it and seeking only evidence that conformed to his theory, once he began to converge on the hypotheses that he wished to support. Regardless of how one judges Freud's own follow-through, his works contain a prototype for studying human thought and feeling—our subjective experience—rigorously. By *rigorously*, I mean in a way that goes beyond interpreting observed behavior and subjects interpretations to an evaluative process that determines their plausibility. The end to be hoped for in the application of this process is not arrival at a final, unequivocal truth, but progress, in the sense of an ever deeper and more complex understanding of the human mind. Human psychology, especially the nature of thought and feeling, is inherently open-ended.

This book demonstrates by example the application of this method of studying mental life through the critical analysis of naturally occurring puzzles that arise in mental life. The book pursues a puzzle that occurs in ordinary mental life. Often the ordinary attracts little notice and stands instead as the implicit, if not explicit, reference point for inquiries into the aberrant. One measures the deviant against the supposedly normal, the irrational against the supposedly rational. Although it is largely uncharted, the ordinary may prove just as elusive as do our foibles. Its puzzles provide a means through which to begin to understand it.

The aim of the book is to solve the puzzle, insofar as it admits solution, and observe in the process the properties of human mental life that are revealed by this effort. In this venture, I have no special commitment to Freud's theory or more specific claims. His claims, like any others, may prove useful or not, or compelling or not, depending upon analysis. I build upon his method.

In the second lecture of his *Introductory Lectures on Psychoanalysis*, Freud explains how sometimes it may be the most trivial behaviors, such as a slip of a tongue, rather than apparent marvels of mental life such as hallucinations and persecutory complexes, that lead to major discoveries about the mind.[9] A small behavior lies at the center of the present investigation. It was first isolated for study by Freud.

# Chapter One

# *The Problem*

In 1904, Freud, then 48, joined his younger brother for a holiday trip to Corfu. Stopping in Trieste on the way, they visited an acquaintance who advised that they would find Corfu too hot and suggested that they go to Athens instead. A boat was sailing for Athens later that day, and it could bring them back in 3 days' time. In an age more immersed in classical civilization than ours is, the prospects of such a trip can only have attracted the brothers. The trip must have tempted Freud in particular, who both pursued antiquities as a hobby and incorporated classical themes and metaphors—the "Oedipal" complex, for example—in his work. Nonetheless, the brothers found themselves dispirited about the plan. In the hours before the ticket office opened, they thought only of the impracticality of the trip and the obstacles that would prevent it. When the ticket office opened, however, they proceeded without ado to book the trip as though they had never questioned it and duly set sail for Athens.

On the afternoon of their arrival in the city, Freud ascended the Acropolis and gazed upon its sights. A surprising thought, as he calls it, suddenly occurred to him: "So all this really *does* exist, just as we learned at school!"[1] (Original German: *Also existiert das alles wirklich so wie wir es auf der Schule gelernt haben?!*[2]).

On the one hand, he was to remark later, he seemed astonished to discover confirmation of a previously doubtful reality. He had responded, he thought, as someone might who had just seen the Loch Ness monster creeping along the shore of the lake: "So the sea-serpent really *does* exist that we've never believed in!"[3] On the other hand, that the real existence of the Acropolis had ever been in doubt was itself a surprise. Freud *hadn't* ever doubted that the Acropolis exists.

Puzzled over the years by this paradoxical thought, Freud attempted to analyze it some 32 years later in 1936, when he was 80. He presented the analysis, a brief occasional piece entitled "A Disturbance of Memory on the Acropolis,"[4] in an open letter to his friend, the French novelist Romain

Rolland. Freud felt guilty, according to the analysis, for being in Athens. His family had been poor, and his parents uneducated. They would have lacked both the interest in embarking upon such a journey and the means for undertaking it. We all feel guilty, according to Freud, when we surpass our parents; he felt guilty for surpassing his. As a result of his guilt on this occasion, he suggests, he attempted to deny that he had arrived on the Acropolis, by imagining that the Acropolis was not there but displacing this impression into the past. He refuted this impression based upon the evidence before him: The Acropolis surely does exist.

Freud's analysis aside, many people experience the kind of paradoxical surprise that he describes when they encounter for the first time objects, places, or events that they have known about but not seen. "Not just a piece of literary self-indulgence after all," remarks a character in Penelope Lively's *Moon Tiger*, when he sees the Sphinx.[5] On a more mundane level, imagine that you have read in your local newspaper of a lamppost that was toppled by a recent storm. Passing by the downed pole while on other errands, you might find yourself musing something along the lines of "So it really did come down . . . ."

One may experience this kind of surprise also when one encounters fresh evidence of the object as opposed to the object in itself. Shortly before my husband and I departed for a hiking holiday in a mountainous region of which we had recently learned, friends showed us pictures of the district that they had taken on a trip there 10 years earlier. Observing the close similarity between these pictures and the ones that I had seen in the guidebook that inspired our trip, I found myself thinking, "It's *real*."

The surprise that people express in these cases is no less paradoxical than it was in Freud's case. Lively's character knew full well that the Sphinx exists and had suggested going to see it. You, on your imaginary errand, harbored no doubts about the accuracy of the news report of the lamppost. My husband and I had bought our plane tickets, booked our lodgings, and mapped out likely hiking trails for our trip. Yet upon seeing the Sphinx, Lively's character responded as though he had doubted its existence. You behaved as though you questioned the authenticity of the report of the lamppost. I responded as though I was learning for the first time that our vacation spot was real.

I shall label as "Freud's thought," or "the thought," experiences of this kind, in which in the face of supposedly relevant evidence people profess surprise at the existence of an object of whose existence they were certain. Unless I indicate otherwise, I shall mean by these terms the general experience, as opposed to Freud's own specific one.

Why would ordinary people, fully accepting of the existence of an object, experience surprise at that object's existence when they saw either the object itself or further evidence of it? The circumstance might remind one of other "tricks" that our minds can play. Sometimes people falsely remember events that they never actually experienced or facts that they never actually heard or read.[6] Also, people experience *déjà vu*, the sense that they have encountered a setting or an object before when they know that they could not have done so.[7]

These apparent misfirings of the mind invite interest because they involve more than mere uncertainty. In a case of false memory, for example, rather than simply wonder whether we put grandmother's watch in the safe as we intended to do, we actually remember having put it there—when we did not do so. In a case of déjà vu, we do not merely reflect that a scene that we have never observed is familiar in some way. We feel that we have been there, and yet acknowledge at the same time that we could never have seen it.

But "Freud's thought" is the opposite of these other experiences. In a case of false memory, people recall something that never occurred. People who have Freud's thought behave as if something that they *did* correctly recall as real were illusory. In a case of déjà vu, one experiences as familiar an object or event that one never experienced. People who have Freud's thought find the familiar strange.

Although the sources of false memory and déjà vu remain controversial, one can envision ways to explain them. Someone's false memory might have been motivated by wishful thinking, for example, or facilitated by the availability of competing cues from other memories.[8] In a case of déjà vu, people's false impression of familiarity of a situation derives from memories of some kind—of past perceptions, emotions, fantasies, and the like—only not of the situation in question.[9] Why people should find themselves surprised to encounter the fully expected, as they do when they have Freud's thought, is less clear.

It stands to reason, therefore, that Freud should have seen fit to classify his own experience of the thought as a momentary quasipathology: an irrational structure that he generated in his own defense (against his guilt for being in Athens). The thought could be seen to parallel a family of pathological disturbances in people's perception of reality that center about the disturbance known as *depersonalization disorder*,[10] in which people experience parts of themselves—their thoughts, feelings, or parts of their bodies—as strange. Closely related to depersonalization is *derealization*,[11] the sense that the external world is unreal. Freud classified his experience on the Acropolis as a case of "derealization" (see my further elaboration in Chapter 2).

One has to wonder, however, as I shall argue more fully later, whether the thought is appropriately characterized as an aberration of any of these kinds. It may align neither with relatively mild "mistakes" such as false memory or (nonpathological) déjà vu nor with outright pathological varieties of depersonalization and its associated disorders. Among other distinguishing characteristics, it resolves into an assertion: that the object in question really exists after all. None of the other conditions that I have described *inherently* resolves into an assertion, let alone into an assertion that rebuts the supposed mistake in question. Freud's thought asserts that the object is real and thus rebuts any notion that the object is not real.

Freud's thought, I shall suggest, is related to a different, ordinary realm of experience. Even if they do not experience the "thought," people are naturally and universally drawn to the real version of things. They gravitate to scenes of events that are reported in the media and to historical sites and relics. They like to see firsthand natural wonders such as total solar eclipses or their child's college diploma. Like Freud's thought, this fascination with the real is less straightforward than it might appear.

People enjoy seeing the real version of things even when they know that the events that occurred indeed occurred and have no question but that the object or the site exists. They want to see the real thing even when they could see pictures and in some cases full-scale replicas of the object or site in question. "I've heard enough about it, and I've seen enough pictures of it, but I hadn't seen it in person," commented one of the hundreds of spectators who in 1995 converged upon the site of the federal building in Oklahoma City in the wake of the unthinkable bombing that occurred there.[12]

People clamor, moreover, to see not only sites and relics of real events that occurred but settings of utter fictions. Japanese tourists fly thousands of miles, and at considerable expense, to Prince Edward Island, Canada, to visit the setting of Lucy Maude Montgomery's *Anne of Green Gables*, which has attracted a following in Japan.[13]

As was the case with Freud, the visitors to these different sites are not (or in the case of Oklahoma City, were not) resolving any doubts about the reality of the objects and events in question. Traveling from hundreds of miles away, cameras and children in hand, the visitors to Oklahoma City cannot have doubted where they were going. The "Anne" enthusiasts do not expect their beloved characters to appear.

People who find themselves moved by these contacts also are not necessarily responding to any perceptual differences that they may detect between the thing in itself and their previous images of it. When a statue at the French Embassy in New York was determined almost certainly to be an authentic Michelangelo, people gazed with awe and disbelief at the very same object that they had observed with less passion previously.[14] Insofar as people were responding to the self-same object, their awe and disbelief could not have been elicited by any change in what they were seeing.

To be sure, one's contact with the real thing can reveal details that one could perceive in no other way. In her middle adulthood, Naomi Fein, writer, legal assistant, and daughter of a schoolteacher, visited the courthouse in Dayton, Tennessee, in which the infamous Scopes "monkey trial" took place in 1925. John Scopes, prosecuted by William Jennings Bryan and defended by Clarence Darrow, was tried for teaching evolutionary theory in the local schools. Well versed in the case since her childhood, Fein noted, among other things, aspects of the approach to the town and of the town itself that "[n]either play nor movie" could suggest.

She also describes her "excitement that edged into anxiety" as she drove toward the area and how she suddenly felt "foolish, emotionally flat" just before she reached the courtroom itself.[15] The physical detail that she noticed did not precipitate her emotional response to her visiting the site. Rather, her emotional response to visiting the site sensitized her to the physical details.

The reasons for people's fascination with the real probably vary. In some cases, for instance on a first visit to the Grand Canyon, one may indeed find oneself prepossessed by the physical display, by a grandeur and a scale that one could not have envisioned through secondhand sources, including photos and films. At least on some occasions, however, as I shall argue more fully in Chapter 9, this fascination with the real becomes obscure in the way that the paradoxical "thought" does—that all this "really *does* exist" when one had no doubt.

In a passing reference to his thought 10 years before he wrote his paper about it, Freud suggests that his experience may have been idiosyncratic.[16] Were it not for his particular circumstances, he says, he might have ascribed his thought to the fragility of the beliefs that he had acquired as a child learning about faraway places. One might infer that this is how he would have accounted for others' experience of similar thoughts, although he does not at any point discuss others' experience.

The latter account makes the thought seem fairly straightforward. The account might be seen to suggest that one's childhood uncertainty about a place resurfaces when one finally visits the place; one then responds to this uncertainty.

Whether or not Freud either intended or would have accepted this implication in the general case, it furnishes an inadequate account of the (general) thought, as I shall show. The common, *non*idiosyncratic version of this experience, despite its straightforward appearance, proves to be even more elusive than Freud's experience was to him. The problem of accounting for it cuts deep, so deep that its meaning ultimately remains somewhat elusive.

I devote this book, therefore, to the pursuit of an explanation of this thought, as a common occurrence, and to the illumination of those more general aspects of our mental life that diagnosis of the thought may reveal.

I shall be seeking what I call the psychological context of the thought. By *psychological context,* which I shall abbreviate as *context,* I mean the family (or families) of thoughts, feelings, or other impressions in relation to which the thought would make sense, from which it would flow naturally. When Freud sought to understand neurotic symptoms and other anomalies, including his "thought," that arise in mental life, he sought their psychological context as I have just defined it.[17]

Not all behavior may have a sense, or at least not a sense that is worth pursuing. One might wonder, for example, whether Freud's thought could be an accidental happening, merely a passing thought with no systematic connection to a context and no interesting meaning to recover. Arguing against this null hypothesis are the facts that the thought recurs and has a definite linguistic shape, or at least an approximate one, whether or not it is spoken aloud. The thought also occurs, as Freud's own thought did, when one is intently focused upon its object, upon which the thought reflects.[18]

The same arguments also undermine the related concern that the thought's context may be too obscure to retrieve. Some observers despair of detecting any coherent meaning in manifestly psychotic irruptions, for example. Yet others insist that even in these limiting cases meaning can be found in principle, if not in practice.[19] Because of its recurrence and intent focus, Freud's thought does not strike the ordinary perceiver as obscure in the sense in which people find wholly dissociated thought to be obscure. One finds the reverse problem. The thought *seems* ordinary and straightforward, perhaps even trivial. As I have warned, it is not so. Hence, I close the chapter with an account of the guidelines that will govern the search for the thought's context.

I shall be attempting to establish a possible context for the thought, as opposed to the actual one in any given case. One cannot know what in fact transpires in the minds of individual people who have the thought. The challenge raised initially by Freud's paper, however, is to discover *any* psychological context from which the thought would flow naturally and without gaps.

I shall, however, require of this context that it be plausible, in the sense that a normal person arriving on the scene could have the thoughts, feelings, or impressions designated as the context.

I lay great stress upon the wording of the thought and shall insist throughout the investigation that the context that we are in search of account adequately for this wording. In stressing the thought's wording, I do not assume that the express thought constitutes the sum total of the experience of the person who has the thought. The person may have noticed, and may have felt, more than is packaged for the thought. Nor do I deny that different uses of the same word string in general may bear only a family resemblance to one another, rather than an identity on account of which all uses share a single, defining group of properties.[20] As it is said, two different people who shout "Brick!" may mean different things by it. One may mean "Watch out! A brick is about to fall!" Another may mean "I've got a brick!"

Rather, the wording of the thought—the expression of astonishment about the *existence* of the object against a background of former *uncertainty* (when one had no uncertainty) about its existence—is curious. Regardless of the experience that the person may be having beyond the words, and regardless of the individual variability that might exist in the precise meaning of the word string, one common question threads through all uses: Why might someone who has never doubted the existence of a given object, and who even sets out with the intention of seeing it, find him- or herself surprised that it exists when he or she sees it (or fresh evidence of it)? To find even one satisfying answer to this question might illuminate facets of the experience, as well as other experiences, that lie beyond the words.

I shall use as cases in point for reflection on this question either recorded instances of the thought, such as those that appear unsolicited in news and feature stories and other published materials, or experiences that people have recounted to me directly. The thought arises sporadically and unpremeditatedly in people. It culminates an experience—a feeling—that wells up and disappears. Although an experience of ordinary people, it probably occurs relatively infrequently. It therefore would prove difficult to elicit experimentally or to observe firsthand on any systematic basis.

I can only speculate about the thought's demographics. Given its impracticality—it expresses one's surprise at the existence of an object whose existence one does not, for most intents and purposes, find surprising—it seems to be a luxury, cognitively. One might expect it to arise, therefore, in people who have the latitude to reflect on their experience (hence among the relatively comfortable), and one might expect it to have evolved fairly recently culturally. Consistent with the latter expectation, an authority in Western classics whom I consulted could not think of any ancient equivalent. The ancients concerned themselves, rather, with the opposite: "Believe it or not, this really exists," when the object in question was truly baffling, such as a marvel or a paradox.[21]

I evaluate candidate contexts for the thought by argument and example, in somewhat the manner of a modern philosophical or linguistic investigation. The evaluation usually appeals to ordinary intuition about clear cases of relevant behavior. The exact procedures of evaluation that I use vary and are best demonstrated by example in the text. In general I attempt to deduce what the behavioral consequences would be if the context in question explained the thought. I then appraise whether those consequences are, or could be, observed. I use this process, in its various incarnations, to work my way eliminatively toward increasingly satisfying accounts of the thought.

In following this overall strategy, I let the problem dictate the course of the inquiry. Existing literature is interpolated along the way, when it is directly relevant to the matter at hand. I do not consider the existing commentary on Freud's 1936 paper. That commentary does not deal with the logical problem that the thought poses. Addressed, moreover, mostly to Freud's own experience, it consists largely of elaborations from a psychoanalytic viewpoint and does not, by contrast with Freud's own paper, consider alternative approaches to the problem.[22]

Before I turn to the analysis proper, in Chapter 2 I give a brief synopsis with minimal reconstruction of Freud's account of his experience of the thought. Although he intended to account only for his experience, the problem begins with his formulation.

The first series of analyses, which begins with Chapter 3, considers, in turn, commonsensical accounts of the thought along the lines of hypotheses that Freud also raises in the 1936 paper (Chapters 3 and 4); Freud's psychodynamic account as a prototype for a more general explanation of the thought, followed by a modification of the account (Chapters 5 and 6); and the hypothesis that the thought might embody the nondefensive, nonregressive revival of a childlike frame of mind (Chapters 7 and 8).

A second tier of analysis begins with a fairly direct expansion of the last hypothesis, to account for the thought's particular emphasis on the real (Chapter 9). Following the realization that children, ironically, would be unlikely to have the thought, I attempt to identify a context for the thought that would accommodate this restriction (Chapter 10). In Chapter 11 I consider general implications of the persisting paradoxical character of the thought and possible manifestations of that characteristic in other behavior.

# Chapter Two

# A Synopsis of Freud's Account

The interested reader may find the English translation of Freud's 1936 paper in the appendix. I continue here at the juncture at which I interrupted Freud's narrative at the beginning of Chapter 1. Having noted the thought's puzzling nature, Freud devotes the remainder of his paper to the thought's explanation.

He considers briefly and rejects the hypothesis that the thought expressed what he describes as the "uninteresting commonplace"[1] that seeing something is different from hearing or reading about it. He sees no reason why an uninteresting commonplace would receive such strange expression. He also rejects the ostensibly deeper alternative that, although as a schoolboy he thought that he believed in the historical reality of Athens, he unconsciously doubted that reality. He finds this account beyond proof and questionable on theoretical grounds.

He finds evidence of a more plausible interpretation in the odd depression regarding the trip that he and his brother experienced in Trieste, before they purchased tickets to Athens. Like the thought itself, the depression in Trieste was surprising, Freud says, given that he and his brother proceeded automatically to the ticket office when it opened. As did the thought, the depression, Freud suggests, expressed incredulity: "It's out of the question that we could see Athens! It would be far too difficult!" Both the thought and the depression, Freud extrapolates, involved his experiencing an event as "too good to be true," as one does, for example, when one hears that one has won a prize or is being sought by the partner of one's dreams.

He takes as self-evident that this kind of incredulity attempts to repudiate a piece of reality. Normally people repudiate realities that threaten unpleasure, not ones that promise pleasure. In this case, the reality in question—a visit to Athens—promised pleasure. Sometimes, however, people fall ill in precisely such situations, when a long-held wish of theirs is fulfilled, as occurs in cases of people who are (pathologically) "wrecked by success."[2] Nowadays, professionals still recognize a "fear of success."[3] Granting that people usually fall ill from frustration, from the *non*fulfillment of some wish, Freud infers that in these cases, in which people's wishes are fulfilled, the external frustration must be being replaced by an internal one. The people cannot allow themselves the fulfillment of their wish, as a result of some deep-seated guilt or fear.

In a similar fashion, Freud hypothesizes, both his depression in Trieste and his thought on the Acropolis attempted to repudiate his long-standing desire to travel and to see the world. Unobstructed, he believes, his thought would have proceeded: "I could really not have imagined it possible that I should ever be granted the sight of Athens with my own eyes—as is now indubitably the case!"[4]

While retaining the incredulity of this supposedly underlying thought, his express thought ("So all this really *does* exist . . . !"), he suggests, displaced this incredulity in two ways. The express thought implied that his incredulity existed in the past rather than in the present, and it attached the incredulity to the existence of the Acropolis, rather than to his standing on it. At some time in the past, the express thought asserted, Freud doubted whether the Acropolis exists. His memory, he says, rejected this doubt as both untrue and impossible.

Freud suggests that these two displacements occurred for different reasons. He shifted his doubt into the past because he had a feeling of the unbelievable and unreal in the present; he could not account for this feeling, given that his senses perceived his surroundings as real. He recalled, however, that he had doubted *something* about the Acropolis in the past, namely whether he would ever see it. This memory provided him with a way to shift his confusing contemporary doubt into the past. The second displacement, on account of which he alluded to a doubt about the Acropolis' existence, Freud thinks, proves that the thought did indeed respond to a current doubt. Given that he never doubted the Acropolis' existence in the past, the doubt of its existence could not have come from the past. He was experiencing instead a "derealization," a feeling that the things around him were not real.

Like their counterpart, "depersonalizations," in which one feels that a piece of one's own self is strange to one, derealizations, Freud says, serve the purpose of defense. They aim to keep something away from the ego, to disavow it. The motive for this disavowal lies in the past, in the ego's store of memories, including distressing ones that may since have become repressed.

The derealization that he believes he experienced on the Acropolis, Freud suggests, repudiated his fulfillment of his long-standing wish to travel and to see such places as Athens. These attainments implicitly denigrated his family and the conditions of his youth. Not only too poor to travel, Freud's father, who worked in business and had no secondary education, would not have appreciated even the idea of going to Athens. The appeal of travel in general, he thinks, may trace to people's early wishes to escape the confining conditions of their youth.

Everyone believes, Freud continues, that it is wrong and forbidden to criticize or to exceed in any other way one's father (the reader may generalize to "parents"). It is wrong to undervalue the very figures whom, still earlier in one's youth, one overvalued as almighty and godlike. The essence of success, Freud speculates, may consist in one's sense of having gotten further than one's father, coupled with the feeling that to exceed one's father is still forbidden. Hence the need to repudiate such success.[5]

I shall evaluate the substance of Freud's account in Chapter 5. First I examine the premise upon which the account rests: that relatively more superficial, or as I shall call them "common-sense" or "face-value," readings of the thought do not suffice to account for it.

# Chapter Three

# *Freud's Thought at Face Value*

To the ordinary observer, Freud's thought sounds as though it has a commonsensical, or face-value, meaning. It expresses, for example, one's awe at what one sees, or, contra Freud, one's observation of the difference between seeing something and hearing or reading about it. In the case of the latter meaning, I refer for now to relatively superficial, for instance perceptual, differences that one may notice between seeing something firsthand and experiencing it through secondhand sources. According to this common-sense understanding, to see, touch, and feel a place in its full, multisensory detail is not the same thing as consulting only an impression of it based upon verbal description or pictures.

By a commonsensical or face-value meaning (viz., context) of the thought, I understand an idea that is expressed more or less directly by the words of the thought ("So all this really *does* exist ... ") that either the ordinary observer or the person who has the thought would retrieve easily in the circumstances. This kind of meaning is to be contrasted with a meaning along the lines of Freud's account, which stipulates the involvement of deeper determinants: concealed, historically rooted impressions and impulses.

A greater number of common-sense readings of the thought exist than Freud considers. One or more of them might account for the thought, as it occurs in general or even as it occurred to Freud. Because these readings appear intuitively to be correct, and because they are the simplest ones imaginable, this chapter and the next contain assessments of them.

Although Freud treats commonsensical readings of the thought too scantily for our purposes, the brief analysis that he provides suggests a model for how to evaluate additional readings of this kind: Envision the wording to which the proposed context would correspond most directly and then ask whether this wording is a good translation of the thought. On this model, if the thought registered one's appreciation of the (e.g., perceptual) difference between seeing something and hearing or reading about it—the "common-sense" hypothesis that Freud considered—then we might expect the thought to manifest in an assertion along the lines of "This looks [feels, smells, etc.] different from what I imagined before." This assertion is not Freud's thought. Hence the proposed context is inadequate to account for the thought.

I shall conduct quasi-experiments of this and related kinds throughout this chapter. Insofar as difficulties emerge in the fit of context to thought, we shall want to know whether the thought eludes (commonsensical) context finding any more than do other ordinary thoughts. I shall conduct this additional comparison separately, in section 2.

In this chapter I ask whether commonsensical contexts can account for the thought literally, and in the next chapter whether they might account for it as a figurative expression. In both cases I mean "account for" in the absence of deeper determinants.

## 1. Freud's Thought as a Literal Expression

In this and the next few chapters of the book, I shall envision the thought to occur, as Freud's own thought did, in response to one's seeing for the first time a place or object that one has heard about but never seen. I shall track as canonical instances of this occurrence the generalized version of Freud's example: a visit to the Acropolis when one cares about the visit, whether or not with the degree of interest and enthusiasm that Freud had.

This example captures the setting of the thought in which most people imagine it to occur, namely one's confrontation with famous, historical objects. It also affords a strong test of the hypothesis that the thought eludes common-sense interpretation, insofar as the setting, with its mystique, invites common-sense interpretation. One imagines, for instance, that the mystique of the object prompts incredulity about it and that the incredulity, combined with direct confrontation with the thing in itself, prompts the thought.

I have added the constraint that the person care about the encounter so as to exclude cynical occurrences of the thought, for instance (the equivalent of): "Ha, ha, so it really does exist [in case some jerk thought it didn't]." Cynical occurrences, of which the example of Lively's that I quoted in Chapter 1 might be read as a case, lend themselves to common-sense interpretation. One has interpolated a figure who could doubt the object's existence. More puzzling, and of interest here, are cases in which people who were certain of the object's existence in all seriousness exclaim to their surprise that the object really exists.

To explore exhaustively whether any common-sense context exists that could explain the thought would be impossible, and the exploration of a large set of such contexts would be tedious. I present, therefore, a sampling of three types of contexts into which the possibilities seem to divide. Because the sampling is still moderately extensive and the treatment of the different contexts superficial by comparison with later analyses, the analysis may appear somewhat dry. It is not technically difficult, however, and provides the foundation for the analyses that follow.

## 1.1. Contexts that Assume No Doubt Whatever About the Existence of the Object

On a completely literal reading, it is incoherent for a person who has had no question about whether an object exists to exclaim upon seeing it, "So it really *does* exist . . . !" (i.e., contrary to what I thought). One cannot both have and not have a question about whether the object exists.

Contexts are conceivable, however, that might appear to justify the thought, while simultaneously making it unnecessary to attribute to the person any doubt about the object's existence. One such context is awe, which I provisionally understand to involve a sense of being overwhelmed by an object for which one feels reverent admiration and almost fear.

Perhaps, then, people reach the top of the Acropolis and are awed by what they see. "So all this really *does* exist," they think, "just as we learned at school!" But even if they feel awe, this is not what the thought expresses. Awe can be expressed in numerous ways, for instance, "Oh!" "Fantastic!" "Amazing!" The thought affirms the reality of the Acropolis.

We might interpolate that people are awed *because* the Acropolis exists. But the thought also does not express this idea. Awe that something exists might be expressed more exactly as "What a miracle! All this exists!" The thought affirms that all this exists *after all*. It alludes to a prior feeling or belief that this does not, or might not, exist.

Perhaps, then, people who have the thought were awed in the *past* that the Acropolis exists. Before they ever set eyes on it, when they knew about it only through things that they had heard or read, they found it astonishing that such a place exists. Many marvel that the Acropolis ever existed or that the civilization that created it ever evolved. The thought might affirm, therefore, that although one previously found it astonishing that such a place exists, it really does exist after all.

Granting people's probable wonderment, we must ask in what this wonderment might have consisted such that it would result, on site, in people's surprised affirmation that the Acropolis really exists.

One reading is that people had some doubt previously about whether the Acropolis exists. We have already discounted this possibility, as did Freud. He never doubted that the Acropolis exists.

On another reading, people can find it awesome (e.g., astonishing, miraculous) that something exists without at the same time questioning its existence. Although they might, for instance, find it difficult to believe that the Acropolis was ever built, they do not now, and did not previously, doubt that it exists. But if they did not at any time question the Acropolis' existence, then it remains obscure why they would be led to affirm its existence when they saw it.

On the other hand, perhaps when confronted with the site people tell themselves that they *could not* have believed that it exists (because it is so awesome, etc.), not that they did not believe it. In other words, the "context" of the thought, the impression that prompts it, consists not of a real but of a hypothetical doubt about whether the Acropolis exists.

This alternative creates a linguistic anomaly. According to this account, the full thought would run: "I couldn't have believed that all this exists. *But* it really does exist." Freud's thought contains no "but." It begins, "*So* [all this really does exist]." One does not answer "I couldn't have believed all this exists" with "So all this really does exist."

Substantively, the problem with this interpretation is that Freud's thought does not allude to a hypothetical doubt ("I couldn't have believed ... "; "It's as if I didn't believe ... "). It alludes to a real one.

An alternative is conceivable that incorporates the attribution to the person of a real doubt, but not about the object of the thought. According to this alternative, the thought responds to one's generic skepticism about one's experience. We cannot always trust the things that people or books say. Or we may call into question our own power to know, remember, or judge, given the possibility of false memory, for example (see Chapter 1). Along these lines, Bertrand Russell wrote that we can never be as sure of those things that we know through indirect means as we can be of those things that we know by direct acquaintance.[1] These observations are not to the point. Russell's idea suggests that, *given* some doubt or question about the existence of the Acropolis, contact with the actual site justifies the conclusion that the place really exists. Freud had no such question; nor has the generic observer had one.

To put the same point slightly differently, when people have the thought, they are not affirming the dependability of things, or of their experience, in general, but of this thing in particular. To recur to one of Freud's supporting examples, were you to find yourself gaping in disbelief at the Loch Ness monster, you would not be registering your surprise that legends in general may be trustworthy. You would be stupefied that the legend regarding this particular object, a legend that you refused to believe, is in fact true.

Any of the impressions that I have described in this subsection—awe, incredulity that the object in question could exist, an awareness of the fallibility of knowledge—could arise as you stand face to face with an object of whose existence you had no doubt. None of these impressions illuminates why you might think: "It's *real!*" or "It's real after all ... ."

## 1.2. Contexts that Assume an Open Question
## About the Object

We do not appear to be able to account for Freud's thought (at face value) by removing from the person all doubt whatsoever about the object's existence. We might fare better by allowing that the person have had at least a minor doubt or open question about the existence of *something* pertaining to the object.

For instance, when Freud saw the Acropolis, he saw something whose exact detail and ambience he could not have anticipated. In the language of contemporary cognitive psychology, one might say that he increased his "depth of processing" of the stimulus.[2] Insofar as he now saw something that he had not seen before, it follows that he could not have believed that *this* Acropolis exists, because he could not have known this object, sight unseen. Therefore, when he saw the Acropolis, he found himself incredulous that the Acropolis as defined by this total configuration of attributes exists.

In affirming, however, that a given entity exists *after all*, the thought refers to something that had to have been the object of Freud's thought previously. Insofar as he was struck instead by something whose nature he could not have imagined (the Acropolis as it now appeared to him face to face), he was not responding to any previously defined object. The appropriate expression of this response would have been something along the lines of "So *this* is what exists!" or "I was unprepared for *this*!" These expressions are not Freud's thought.

We could, however, allow that the thought might have responded to this more complex object by allowing *that object* to have been the object of Freud's previous thoughts. He had read about or seen in pictures an object that struck him as a fantastic place or that he understood to have a given layout, architecture, and other details. Either of *these* objects could have occupied his reflection prior to his seeing the object firsthand, and he could have questioned the existence of either one. He could have wondered whether *so fantastic a place*, or *a place that looks like that*, exists. Standing there, he could affirm that it did.

Nonetheless, these antecedents also, except by a sleight of hand, do not address the question of whether the Acropolis exists but the question of what it is like: "Is it as fantastic as I imagine it to be?" "Does it look like that?" Even if one wondered whether "so fantastic a place" (etc.) exists on the site of the Acropolis, one would be unlikely to believe that *nothing* exists there. One would still think that something exists there that contains ancient Greek remains. Otherwise, why go there? Further, if having arrived on the Acropolis for the first time and exclaimed that it really exists, one would be responding oddly if one were referring to anything other than the thing one was standing on. The thought *sounds* as if it refers to the thing in itself, under any description.

A further, rhetorical difficulty arises with the aggregate of interpretations that we have been pursuing in this subsection. First we examined the interpretation that the thought responds to the vast *difference* between seeing the real thing and reading or hearing about it. Next we explored the interpretation that the thought responds to one's discovery that the object indeed looks exactly like one's image of it. Although different people could respond in either of these ways to the same object, one has to wonder whether the same expression is likely to express two such opposite meanings, even on different occasions. In the absence of grounds for favoring one meaning over the other, the line of interpretation as a whole is suspect.

## 1.3. Contexts that Assume an Antecedent
## Doubt About the Existence of the Object

Interpretations that attempt to dilute the thought's allusion to a doubt about the object's existence do not produce the thought. One way remains to salvage face-value interpretation of the thought: Admit in some fashion the stronger allusion—that the person did at one time doubt the object's existence—while preserving the assumption that he or she also has not doubted it. Here are three possible scenarios.

(1) A sense of unreality might have surrounded talk of the Acropolis and of ancient Greece when Freud was a child. To this day, the real is interspersed with the unreal in our lessons. Socrates was real. Achilles was not. Standing as adults on the Acropolis, people might respond to this earlier ambiguous status of the period and its objects.

However, even if people wondered as children about the Acropolis' reality (a condition that Freud denies of himself), as adults they do not doubt it. The adult has the thought.

(2) Perhaps, then, people's child mentality regarding the site simply asserts itself upon their arrival there. But then an explanation would be needed of why it does so. That explanation would exceed what I am calling interpretation at face value.

(3) Nonetheless, even as adults we may seem to feel that some places that we know about are unreal. A mature European who has never traveled to the Western hemisphere may feel that the existence of the Grand Canyon is a myth, depicted only on postcards. A North American who has never traveled to Europe may wonder whether Luxembourg really exists or whether it is just a place on the map. Insofar as one does wonder, Freud's thought might respond to this wonder.

In what sense, though, does one wonder? Officially one does not wonder. If asked, the European would agree that the Grand Canyon exists, and the

North American would agree that Luxembourg exists. Either would expect to see the place in question if presented with an occasion to see it. If Freud's thought did respond to the less official sense of wonder, the questions would remain of why this (unofficial) sense of wonder, rather than the official one, asserts itself on site, and of what this unofficial sense of wonder is.

## 2. Literal Interpretations of Other Thoughts

The foregoing contexts fail to account (literally) for Freud's thought for one or more of three reasons: They undermotivate the thought (section 1.1), suggest the wrong thought when fully elaborated (section 1.2), or sneak in additional layers of interpretation and thereby exceed "face-value" interpretation (section 1.3). One wants to know whether these problems reflect general ambiguities of translation between (purported) meaning and expression in the case of other assertions or whether they reflect a difficulty with Freud's thought in particular.

The latter seems likely. It is easy to invent for other thoughts (that are to be understood literally) plausible contexts that map onto the thoughts with the degree of precision that I was seeking in the foregoing analysis of Freud's thought.

Suppose that Freud (whom I use for now as a generic actor) had had the thought "I think the tickets are in my back pocket." It is easy to envision (commonsensical) contexts in relation to which he might have had this thought. He might have been wondering which pocket his tickets were in: shirt pocket or back pocket? He might have been wondering—perish the thought—whether the tickets were in his back pocket or whether he had left them home. He might have been musing that he had forgotten tickets in the past or that in general this is something that people have been known to do. Or, with no clear antecedent thought, he might just have needed to reassure himself about the tickets.

It is easy to suggest contexts for this thought because it addresses a doubt or question and can be accounted for by reference to that doubt or question. Not all thoughts presuppose a doubt or question, yet we can conceive possible contexts for them.

The declaration "The sky is blue" is trivially obvious, yet one might have reasons to assert it. One might be thinking of obvious things that people sometimes say, or one may just want to state the obvious for some reason. One might be thinking of things that are blue or of colors that standardly appear in a landscape.

Odd though the equally obvious thought "All this really exists!" might seem (when the statement is asserted in reference to the Acropolis as one stands on it), for this thought too we can interpolate a possible context: "The world contains some wonderful things. All this exists!"

Freud's thought asserts the same, obvious claim as does this last thought: The Acropolis exists. It eludes interpretation because it assumes the form of an affirmation: "[So] all this really *does* exist . . . ." We cannot easily invent a context for this affirmation because it presupposes a question about the existence of the object when the person has not questioned its existence.

A parallel elaboration of the assertion "The sky is blue," however, suggests a way in which we might provide a context. One might assert that "the sky really *is* blue" and mean by it not that the sky is blue in case someone doubted it but that the sky (for example, on this particular day) is very, very blue. The segment ". . . really *is* . . . " would mark intensity and emphasis in this case, as the philosopher J. L. Austin[3] suggested that words such as *real* or *really* may do. Correspondingly, Freud's thought might mean that all this exists with a vengeance (as in "It really *is* raining," i.e., it's raining cats and dogs), rather than that it exists, as opposed to not existing.

This reading ignores the allusion of the thought to a previous set of deliberations: "So [all this really does exist] *after all*." It would be incongruous to assert "So all this exists with a vengeance, after all" or "So all this superexists after all." The thought concerns the truth of its predicate (the Acropolis either exists or it does not), not the intensity of the quality that the predicate attributes.

Nonetheless, one can provide a potential context for "So the sky really *is* blue . . . ," given this constraint, that the assertion address the question of *whether* the sky is blue (when no one doubts it). The person could have been questioning the obvious. She might have been wondering whether the sky really is blue or whether its apparent blueness results only from the way in which the light is diffracted through the air. ("The sky is blue. Is that really so? Or does it only look blue because it stretches off to infinity?"[4]) But then, this inquirer might have reasoned further, isn't the color of all things the result only of the way in which the light is diffracted or reflected from various surfaces? So, the sky really *is* blue, in at least the sense in which a rose is red.

Freud's thought could be similarly elaborated to include some kind of abstruse reflection from which the conclusion "So all this really *does* exist!" would follow. For instance: "I can't see it so maybe it isn't there." In the case of Freud's thought, we have additional information that blocks this sort of account. We know from his self-report that Freud did not entertain any skeptical premise and was not recalling one when he had the thought. When other people have it also, the thought wells up suddenly and unpremeditatedly, in contrast to the case of a conclusion of a line of conscious reflection.

The literal interpretation of Freud's thought seems to pose a problem that does not arise with the literal interpretation of other expressions.

# Chapter Four

# *Freud's Thought as a Figurative Expression*

The possibility remains that Freud's thought could be a figurative expression, a way of talking, that might assert indirectly one or more of the commonsensical meanings that I considered in the last chapter. It might express awe or one's awareness of the difference between a seen and described object, for example, metaphorically or by some other allusive means. In this way, it would express meanings that "common sense" judges to be the correct ones and would escape the requirement, imposed in the last chapter, that this meaning map directly onto the thought. To assess whether the thought is likely to be an assertion of this kind, in this chapter I compare the thought with different classes of figurative expression.

## 1. Linguistic Tropes

Linguistic tropes, such as metaphor ("The fog comes/On little cat feet...") or idiom ("Let's shoot the breeze"), are assertions that express a meaning that cannot be generated from the literal meanings of the constituent words. Freud's thought does not appear to be an instance of a linguistic trope. It has a meaning that can be generated from the (literal) meanings of the constituent words: All this really does exist, just as Freud, for instance, learned at school.

The thought also appears to fall outside the category of discourse tropes, such as hyperbole, which exaggerates something to the point that it cannot be true: "I could have killed him." "I could have died on the spot." "Pat Buchanan is the most evil person since Hitler." You *could* have killed your associate for spilling the petri dish, but you wouldn't have. You *could* have died on the spot through embarrassment at your latest faux pas, but you wouldn't have. Pat Buchanan could be the most evil person since Hitler, but he isn't, unless you have an odd standard of evil. Freud's thought contains no exaggeration or any other evident distortion. The Acropolis and its surroundings really exist, and Freud—as a sample subject—learned about them in school.

The thought as a whole, on the other hand, might seem to exaggerate the less dramatic idea, that all this exists *in just the way that* it was described in school. It is just as awesome as people said. The thought does not, however, merely exaggerate the idea that the Acropolis as seen looks just like the thing in the textbooks. It states a wholly different idea: that the object exists at all.

## 2. Indirect Discourse

People engage in other indirect acts of speech[1] (and, presumably, thought), however, that although not tropes, allude to a meaning that differs from what the words say literally. On a literal reading, "Could you pass the salt?" implies that (the speaker thinks) you might be incapable of passing it. The speaker intends to discover, however, whether you *would* pass it, not whether you know how to pass it. The express utterance is a circumlocution. The circumlocution has a purpose. As a result of their socialization, people find it uncomfortable to ask others directly to perform favors. Instead they wonder, via such devices as the indirect request, whether the burden on others is not so great that they actually *could* pass the salt, as opposed to not being able to pass it.

Freud's thought might be conceived as a circumlocution of a similar kind, an indirect way of asserting, for instance, that seeing something is different from hearing or reading about it. If the thought, whether spoken or private, did make this allusion, however, the question would arise as to why it did so indirectly. The motive for the circumlocution in the case of passing the salt is clear. "Could" is more polite than is "would."

## 3. Loose Usage

Language may also be used, though, in ways that are loose, rather than indirect. Some of these usages share with Freud's thought the properties that they recur and that they have somewhat obscure meanings.

You disparage a recital that you have given, that I did not hear. I say, in all sincerity, "I'm sure it was fine." No matter what the circumstances, I cannot be *sure*. I wasn't there. Hence I speak loosely.

Freud's thought could be seen to lack precision in the same way that this assurance about the recital does. Both assertions lack background assumptions whose presence they imply on a literal reading. I lack complete certainty regarding the recital, though I say, "I'm sure it was fine," in my judgment of it. People who have Freud's thought lack *un*certainty with respect to the object of the thought, though they affirm the reality of the object, as if some question existed about it.

A pretext exists, however, for the statement "I'm sure [the recital] was fine." I am reassuring you about your performance, responding to your doubts. In the light of these doubts, I am saying I have good grounds for believing that what I assert is correct. Given my knowledge of you, of performers in general, and so forth, I have every reason to believe that the recital was fine. Even if I do not have good grounds for reassuring you, I am asserting at least my uninformed faith. Although Freud's thought that "all this really *does* exist" also suggests the presence of "good grounds" for believing something, no question has been raised about the belief.

Freud's thought, in its setting, resembles more closely the assertion "I'm sure the chair is blue," when one asserts this while looking directly at the chair. Certainty is licensed in this case but would be inappropriate to assert, because no one has any doubt. If you and I could not see the chair's color entirely clearly, then I could insist appropriately, "I'm sure it's blue." But people do not ever assert "I'm sure the chair is blue" when they are looking straight at it and have no doubt about it. They do assert Freud's thought.

To the extent that the thought recurs in the same linguistic form (e.g., "So it really *does* exist . . . !" or "So it *is* real"), it could be seen to share qualities of a (nonidiomatic) cliché: a trite or stereotyped phrase or expression such as "strong as an ox." But if it either is or resembles a cliché, we do not know its meaning. We know neither the meaning around which the supposed cliché took shape originally nor the meanings that the cliché may have come to represent since.

## 4. Summary of Common-Sense Interpretation (Chapters 3 and 4)

Freud's thought resists interpretation at or near face value, as an assertion to be understood either literally or figuratively. On the model of other assertions that may be understood literally, Freud's thought responds to a doubt about whether the Arcropolis exists; one overturns this doubt in the face of the evidence. This reading does not suffice, because the author of the thought has had no such doubt. The lesser questions and surprises that I considered in section 1.2 of Chapter 3 might more plausibly be entertained by a person arriving on the Acropolis. For instance, one does not know exactly what the Acropolis looks like or how one will feel upon seeing it. Relative to the mapping from context to thought in the case of other expressions (such as those presented in section 2 of Chapter 3), however, these lesser questions and surprises do not map onto Freud's thought.

The thought also does not exhibit the properties of typical nonliteral assertions. Other nonliteral expressions either are more obviously grounded in some relevant set of presuppositions (as in the case of "I'm sure [the recital] was fine") than is Freud's thought, or their indirect form is more clearly motivated (as with "Could you pass the salt?") and more transparent (both of these cases, as well as metaphor, idiom, or cliché). One either grasps their meaning immediately, or given sufficient background knowledge (e.g., about clouds and cats' feet, in the example of metaphor that I gave in section 1 of this chapter), one can reconstruct it.

# Chapter Five

# *Freud's Account Analyzed*

The difficulty of face-value interpretation of the thought now apparent, Freud's psychodynamic account of the thought assumes interest. Granting that we do not know whether Freud himself would have generalized his account to other cases, it stands as a kind of account that was designed to avoid the pitfalls of face-value interpretation. This chapter analyzes the account, therefore, to assess whether it, or something like it, might be extended to the general case of the thought.

Freud believed that all human behavior, except for genuinely accidental behavior, which he thought occurred rarely,[1] made sense, within *some* (psychological) context. It flowed naturally from some neighborhood of thoughts, feelings, or other impressions.[2] Insofar as a behavior did not appear to cohere within any commonsensical context, then, Freud reasoned, the behavior might make sense within a context of which the person was not aware. That context would consist of unconscious impressions and impulses rooted in the person's history, but ejected from consciousness in the service of defense. This is how Freud explained both markedly abnormal structures, such as dreams and neurotic symptoms, and momentary aberrations such as slips of the tongue.[3]

Freud analyzed his thought on the Acropolis on this model. His guilt in reaching Athens—guilt of which he was unaware at the time—he suggests, prompted a defensive process within him that resulted in his impression of the things around him as unreal (his "derealization"). His thought that the things around him indeed were real responded to this impression, but displaced[4] it into the past, because of its conflict with his otherwise intact perception.

I consider first the cogency of this account for Freud's own experience of the thought and then the potential applicability of the account to the general case of the thought.

## 1. Freud's Account of His Experience

Freud's account of his experience depends upon three assumptions: that he felt the guilt that he describes; that his thought responded to this guilt; that the thought responded to this guilt through the mechanisms that he describes.

Freud builds a plausible case for the first assumption, that he felt guilty. In addition to explaining why he might have felt guilty in Athens—he sensed that he had surpassed his father by being there—he describes several signs of his unease. These include the anticipatory depression that he and his brother experienced in Trieste, before they headed for Athens. In addition, he reports both his experience of a "derealization" on the Acropolis, which provides potential evidence of guilt or other unease, and a "reserve", that he says both he and his brother shared, surrounding the whole episode.[5]

That the thought might have responded to this guilt—the second assumption of the account—is supported by the fact that Freud's affirmation of the Acropolis' existence otherwise appears to have no context. Although we have not found an adequate context for the thought among the interpretations of common sense, however, a plausible context might be found in some variety of deep interpretation other than a Freudian (that is to say, psychodynamic) one. Freud made no inquiries in this direction. It was not his objective to do so. His aim was to see whether a psychodynamic vantage point could illuminate his thought, given the limitations of common-sense analysis.

Freud's account of how the thought might have responded to his guilt (the third assumption) is curious. The thought, he says, replied to his derealization (which reflected his guilt), but displaced the feeling of unreality into the past, because he could not understand his contemporary feeling of unreality. The feeling retracked specifically as an impression of a past doubt because historically he did doubt something about Athens, namely whether he would ever see it.

It is an odd idea that Freud's contemporary feeling of unreality can have been so disturbing that it motivated a displacement of the feeling at all.[6] Also puzzling is why, insofar as a displacement might have occurred, it would have selected precisely the pathway that Freud was trying to avoid: the issue of whether he would, or ought to, see Athens. Although defensive processes are not supposed to be rational, and they routinely incorporate contradictions, this one seems unusually convoluted, relative to other accounts of Freud's of defensive formations.

Elsewhere, for example, Freud recounts the story of a patient whom he had diagnosed as suffering from delusional jealousy.[7] Desperately afraid of her own attraction to her son-in-law, according to Freud's analysis, the woman created the illusion that her husband was having an affair with a young woman at his factory. With this illusion, Freud surmised, the woman relieved some of her guilt. If her old husband could be having an affair with a young woman, then she need not assail herself for being in love with a young man.

The woman had two impulses, according to Freud's diagnosis. She wanted to deny her (fantasized) infidelity (her attraction to her son-in-law), on the one hand, and she abhorred herself for it, on the other. She satisfied her need to deny her infidelity by reducing her guilt through the allegation against her husband. She satisfied her self-abhorrence by suffering at the thought of her husband's infidelity.

Had the woman's defense followed the pattern of Freud's supposed defense on the Acropolis, not only would she have attempted to satisfy these two contrary impulses. She would have directly undone one with the other. For example, to lessen her distress at her husband's supposed infidelity, she might have recalled her own faint attraction to her son-in-law! "I forgive him [my husband], for I am bad too." Had she done this, she would have negated the whole purpose of the defense. She would have been admitting exactly the impulse—her attraction to her son-in-law—that she was trying to avoid.

Defenses of this sort, in which the idea that one is presumably trying to repress surfaces in some form, do occur, according to Freud. Either the defense as a whole has failed, and the repressed has "returned", or, as a sign of returning health, the person begins to admit repressed material into consciousness.[8] Freud gives no indication that a process of either of these kinds was occurring when he had his thought. The whole idea of a tortured battle between two contrary aims—the need to deny his arrival, on the one hand, and the need for perceptual intelligibility on the other—suggests an overly dramatic construal of the circumstances, even in Freud's case.

I view these quirks in the account as more strange than devastating; other accounts of Freud's are smoother and more convincing. The account could avoid the quirks if it assumed more simply that in attempting as a result of his guilt to avoid acknowledging that he had indeed reached the Acropolis, Freud exulted instead in the Acropolis' existence. He could exult in the Acropolis' existence "after all", because he had known about it for a long time—after all. Thus, what he could not admit of himself, he projected onto the Acropolis. "Projection" is a familiar defense mechanism in Freudian analyses[9] and many since.

The possibility of a coherent account along these lines notwithstanding, the problem with the second assumption remains. Even if Freud felt guilty for having reached the Acropolis, and the thought *could* have reflected this guilt, did it do so? We cannot know. Plausible alternatives exist. Even if Freud felt a certain "reserve"[10] surrounding his arrival in Athens, the thought also joyously acknowledged his arrival there. It could therefore have represented a lapse in his defense, rather than an extension of it—though we have as yet no notion of how it might have done this.

## 2. The General Case of the Thought

The problems with Freud's account become magnified when one tries to extend the account to the general case. Taking the assumptions of the account in reverse order:

In addition to any technical problems that may arise with the mechanisms through which guilt of the sort that Freud describes would map onto the thought (the third assumption), we now have the problem that the thought recurs. Different people have it. Even if the thought served as a defense whenever it occurs, the question would remain of why this defense is the one that occurs. If everyone who has the thought has it, say, to deny a given point of arrival, there are numerous other evasions among which to choose. One could think about the weather or the alignment of the columns on the Parthenon, or one's plans for lunch. Why choose this defense in particular? Why would *so many* people choose it?

As Freud recognized, others fall prey to the conflict that he describes, between accepting and denying their good fortune, which they, as he did, regard as "too good to be true." Their background would not have predicted, or more strongly would not have allowed, such a result. By contrast with Freud, however, at least some of these other people do not displace this denial.

The Irish poet, Seamus Heaney, winner of the 1995 Nobel Prize in Literature, recounted at the awards banquet how as a young man he had no idea that he would ever visit Stockholm, let alone as a guest of the Royal Swedish Academy and Nobel Foundation. Quoting Oscar Wilde, he said that "'the only way to get rid of a temptation is to yield to it.' All along," he continued, "I have been resisting the temptation to believe that I have actually won the Nobel Prize. Now finally . . . I can yield to it."[11]

Heaney, according to this account, never expected to win the Nobel Prize, had difficulty believing that he had won it, and then, as the ceremonies progressed, gradually coddled himself into believing it. Freud, according to his analysis, never expected to reach Athens and had difficulty believing that he had arrived. "By the evidence of my senses," he describes himself as having felt, "I am now standing on the Acropolis, but I cannot believe it."[12] Then, however, he affirmed not that he had indeed arrived, but that the Acropolis is real. He "displaced" the more direct affirmation (that he had arrived), according to his account, because of the guilt that that affirmation would have aroused.

Heaney's experience, which lacks the supposed displacement, could also be seen as defensive, insofar as he denied the truth.[13] Additionally, as Freud suggests, denial exists in the very idea that a plainly documented fact (such as official notification that one has won the Nobel Prize) is *too* good to be true. A fact is either true or false. Insofar as one feels that it is "too" good to be true, then one must feel undeserving of it for some reason.

One may wonder, therefore, whether, even if the pathway to yet further displacement of one's acknowledgment of one's success lay open, one would take the path. Possibly Freud felt more conflicted than did Heaney and needed to defend himself further. Then the question would remain only whether the thought could arise as the result of such additional displacement.

However (regarding the first and second assumptions of Freud's account), even if Freud felt guilty in Athens, it is difficult to imagine what would motivate others' defensiveness upon their encounter with the real version of things of whose existence they had known. Freud, as we know, remarks that his own circumstance may have been idiosyncratic.[14] If others do not feel guilty, then the thought could not be a defense against guilt.

# Chapter Six

# *A Different Neurosis*

Freud's account still holds the advantage that it explains how the thought might address a doubt (about the object's existence) when in fact the person had no doubt. Applied to the general case, the account says, in effect, that through a defensive process occurring within them, people may generate an irrational doubt. They then respond, via the thought, to this irrational doubt. Although Freud's particular account has problems, some other account along these lines might succeed.

Freud's idea that the impetus to his thought lay in his "derealization" (his feeling that the things around him were not real) indirectly suggests a possibility. That possibility builds upon the prospect that a derealization may still occur in conjunction with the thought, even though this derealization may not, contrary to Freud's proposal, explain the thought. I shall work my way toward this alternative account in this chapter. I begin with a preliminary account that, although inadequate in itself, will serve as a basis for elaborating the account that I wish to entertain more seriously.

## 1. A Preliminary Account

Let us suppose that a feeling of unreality (call it a derealization for convenience) does accompany the thought, as it did in Freud's case. We can imagine that an aura of unreality might surround at least those objects that one sees of which one is in awe, as one might be of the Acropolis.[1] Now, let us suppose that, rather than *elicit* the thought, the feeling of unreality (viz., derealization) occurs in parallel with the thought, prompted by some third impression. What might this antecedent impression be? Consider the following situation, in which both a feeling of unreality and a version of Freud's thought might occur, in parallel.

At the airport, you are seized by the worry that you might not have your plane tickets. You check one pocket, then another, than another, and finally find the tickets in your briefcase. You feel immeasurable relief. After such experiences of high anxiety, when one finds the object that one is seeking, one may for a moment perceive the object to be unreal or may doubt its reality. Having so alarmed oneself, one is not sure that one has really come through. At the same time, one knows that one has the tickets and might be moved to affirm, "So I do have them." The first reaction, the feeling of unreality, prolongs the doubt, as a kind of resonance of it. The second reaction, the recognition that one has the tickets after all, answers the doubt and signals its elimination.

In this case, the thought "So I do have [the tickets] after all" responds not to the feeling that the tickets are not real but to the doubt that one had moments before that one did not have them, plus the evidence that one does have them. This same doubt, plus the advent of the tickets, also causes one's feeling of unreality.

Let us now apply this same scenario to Freud's (own) thought on the Acropolis; I use Freud here as a sample subject. According to his account, his derealization was caused by his feeling that he should not be in Athens and the evidence that he was there. Let us adopt this part of the story too. Following the model of the airport scene, we should now suppose that the thought "So all this really *does* exist . . . !" responded to this same feeling of guilt—the very feeling of guilt that caused the derealization—and the evidence that Freud was in Athens.

This match of provocation to thought is less good than is the match of provocation to thought in the case of the tickets, however. In the case of the tickets, the provocation is the fear that you do not have them. The thought, moments later, is "But then, so you do have them!" The instigation of Freud's thought on the Acropolis, according to his account, was his fear that he should not be there. The answer (i.e., the thought) was that the things that are standing there exist after all. Some additional step is needed to explain how the thought came to assert something about the Acropolis rather than about Freud. By our current assumption, that step was not Freud's derealization. Some other doubt, however, might explain both the thought and the derealization.

## 2. What If . . . ?

Put yourself back at the airport. You check to make sure that you have your tickets. Five minutes later, having moved only a few feet from your original location, you check again. The area is not rife with pickpockets. The tickets would have to have jumped out of your pocket to be missing. Why do you check nonetheless? The consequences of not having the tickets would prove so distressing—missing the flight, facing the chore of obtaining new tickets and a new flight, feeling as though you had been violated or as though you sabotaged yourself—that you want to make sure that you have the tickets.

In these situations, one may be thinking "What *if* [the tickets] weren't there?" The prospect alarms one sufficiently that even in the absence of any original doubt, one may provoke a doubt. At least one might start behaving, and feeling, as one would if one had an authentic doubt.

Freud's thought might have a parallel impetus. (I again use Freud as a sample subject.) Freud arrived on the Acropolis and had the thought "What if it didn't exist, after all?" The consequences would have been severe. His earlier hopes and dreams, his feelings of expectation, and his trip would have been embarrassingly pointless. He would have been a gullible, really stupid, little boy. But he was not, because all this really does exist.

The question "What if . . . ," however, needs a provocation. In the case of the tickets, insofar as you are not looking at them when you start to imagine the consequences of not having them, an infinitesimal possibility exists that they could be gone. (Nonetheless, that one fixes on these infinitesimal possibilities might merit explanation.) When Freud visited the Acropolis, he did not experience any doubt before he arrived on top. As he mounted the hill, he did not wonder whether the things on the Acropolis were really there. Why, then, might he have posed the question "What if it did not exist?"?

Various scenarios are conceivable. He might have thought of other situations in which this wariness would have been warranted. He had been foiled or duped before, or so he might have thought. Or, as per his interpretation, he felt that he did not deserve his attainment. Suddenly to ask "What if it didn't exist?" would at least have accomplished a reduction in his pleasure by wasting his time with a distracting question.

More benign provocations are imaginable. Suppose that you had studied French for several years. Now you go to France. To your pleasant surprise you observe, "They really do speak French here!" Imagine if they did not. Reality would not align with expectation, and all those years of study would have been in vain.

This account furnishes Freud, or the generalized observer, with a doubt, and the doubt is something other than a (real) doubt about whether the Acropolis exists. It compels a response, insofar as it produces a state of mind that feels like a real doubt, which, if it were real, would compel a response. Even if this doubt were motivated, as Freud supposed to be the case in his own incident, by the feeling that one should not be where one is, it concerns the object of the thought and not the subject. It therefore demands a response that refers to the object, as Freud's thought does.

## 3. A Pure Affirmation

The question "What if it didn't exist?" does not flow as smoothly into Freud's thought, however, as the question about the tickets flows into the

affirmation that you have them after all. In the airport, you slip into thinking, "What if I didn't have the tickets?" You check, and then you think,
you do have them after all. On the Acropolis (or in a similar situation),
there is no prior question that prompts a search and then prompts the confirming thought. The confirmation that the Acropolis exists occurs simultaneously with the supposed wondering. Freud was wondering, according
to this account, "What if it didn't exist?" while he was looking at the
Acropolis. To have assured himself that it does exist *after all* would have
been incongruous.

The faithfulness of the account to the subjective reality of the thought
must also be questioned. If on the occasions when one has Freud's thought,
one asks, "What if [the object] didn't exist?" one has no awareness that one
is doing so. One experiences none of the anxiety, tension, fear, and ultimately relief that one feels in situations such as those at the airport.

One does not appear to think anything negative at all. The thought seems
to be purely affirmatory, despite its allusion to a doubt or question. Once in
Athens, you mount the Acropolis. There everything is. It really *does* exist
. . . ! You not only are unaware of any doubt or question but take particular pleasure in your observation and feel release through it. Contrary to
both this account and Freud's, these are not the qualities of defensively driven or otherwise burdened behavior.

# Chapter Seven

# *The Thought as Mundane*

Not only affirmatory and pleasurable, Freud's thought is *mundane*. Given that the person has not doubted the existence of the object of the thought—and that in the case of a well-known object such as the Acropolis, no one else would doubt it either—the thought asserts the obvious. All this that everyone knows exists, exists.

An account that could explain these qualities of the thought—its affirmatory and pleasurable quality, and its mundanity—might be able also to explain its thus far elusive content. I attempt to develop an account of this kind in the next chapter. In this chapter, I elaborate and provide support for the claim that the thought is mundane. To expose this characteristic more fully, I first locate the (canonical) thought on a continuum of assertions that affirm the existence of realities ranging from the miraculous to the mundane. I then adduce some new, noncanonical examples of the thought.

## 1. Miracles and Lesser Events

Consider the most unbelievable sort of event, a miracle, which we may define as an event for which ordinary physical explanation fails to account. Examples of miracles would include a picture crying, a sea parting, or the Loch Ness monster turning out to be real. Although the reality of these events is questioned, the events are spoken of, and when they are, they are described as miracles.

Freud thinks that his thought on the Acropolis expressed at least in part the kind of shock that one would feel upon encountering a miracle. Were one to see the form of the Loch Ness monster creeping along the shore of the lake, one would be forced to admit, he imagines, "So it really *does* exist, the sea serpent we've never believed in!"[1] (*"Also existiert sie wirklich, die Seeschlange, an die wir nicht geglaubt haben!"*)[2]

39

But think. Although one might be "driven to admit"[3] this idea sooner or later, it is not what one would assert on the spot. Imagine actually being there. There's the monster. You would more likely respond, "Oh my god, it's *real!*" or "*Yikes!*" Evidently the hoaxers who were responsible for the 1934 picture of the alleged monster shared this intuition. One of the party attributed to his (fictitious) companion the response, "My God! It's the monster!"[4]

One's response to nicer miracles would be similar. Suppose that Santa appeared in your driveway. You would think, "Oh my God, he's *real!*" or, simply, "Oh, my *God!*" Screenwriters for the film version of Thurber's classic tale "The Secret Life of Walter Mitty"[5] displayed a similar expectation. Upon accidentally discovering a trinket given to him by his true love, whom others had insisted was imaginary, Mitty exclaims in dismay, "She's *real!*" and dashes off to find her.

None of these reactions is Freud's thought. The hoaxers' fictitious observer did not say, "So it *is* real, after all!" Mitty did not exult, "So she's real after all, just as I thought . . . !" These responses, the equivalent of Freud's thought, would be out of place at the scene. When one has Freud's thought, therefore, one neither has encountered a miracle nor thinks that one has.

Nonetheless, people sometimes speak of the existence of such objects as the Acropolis as "miraculous." Insofar as they do so, they speak metaphorically. They mean something more like astonishing, seemingly out of the realm of human capability, or out of the realm of human capability as one perceives it to have existed in antiquity. Yet even this somewhat diluted meaning of *miraculous* may not figure in Freud's thought.

People find various spectacles astonishing and describe them as miraculous, although they do not mean miraculous in its full-blown sense of involving the supernatural. The whole idea of a mummy, for example, is startling: the thing in itself, its function and what it accomplishes, the process that produced it, the fact that it was ever conceived, the fact that it was conceived when it was. Or consider total solar eclipses. Does the sky really turn all black, in the middle of the day? Does one really see a fiery ring around the sun? How could it come about that the sun, moon, and earth ever are aligned so exactly that this sort of thing occurs? At bottom, one believes that mummies exist and that eclipses occur. These questions may cross one's mind nonetheless.

To be sure, the Acropolis is awesome. The physical structures are impressive. That such a place ever existed, that it existed when it did, perhaps eludes our full comprehension. But strictly speaking, we do not find it difficult to believe that the structures that stand on and around the Acropolis today exist.[6]

Consider still smaller scale wonders. You look through a telescope for the first time, at Saturn. "It really *does* have rings!" On the one hand, you didn't doubt it. On the other hand, Saturn *is* out of the ordinary when compared with the other celestial bodies that we see with the naked eye. It is an object of wonder, and we may wonder.[7]

Suppose, on the other hand, that you are on your first transatlantic flight, traveling from the United States. You experience firsthand the vast expanse of ocean extending between the continents. "So the Atlantic Ocean really *does* exist!" you muse. Even if you lived on the East Coast of the United States and visited the beach, where you saw the ocean, this is different. You are experiencing the Atlantic Ocean as an ocean, the very thing that they showed you on the maps at school, filling up all of that space in between the continents.

The Atlantic Ocean is not normally an object of wonder in the way that Saturn is. As the plane takes off, you do not wonder whether the sea will be there. Rather, you have made the quaint observation that you can actually see this entity that "stretches," as they used to say in school, between the two continents.

Freud's thought, as expressed on the Acropolis, concerns a reality that most people would consider to be less obvious than they consider the Atlantic Ocean to be. Yet the thought significantly resembles the thought about the Atlantic Ocean. As you reach the base of the Acropolis, you do not wonder whether the things that are on top are really going to be there. You get to the top. You are in a foreign country (unless you are a resident Greek), in contact with the remains of a culture that existed longer ago than one can fully fathom and that produced a legacy that few can believe. But for a moment, none of that is what you respond to. You observe that all of this—the physical site and the things around it—really does exist, just as you learned at school!

## 2. Further Examples of the Thought

Insofar as Freud's thought observes something mundane, it might occur in mundane settings, and not only in settings that one usually thinks of as awe-inspiring, such as a visit to the Acropolis. In this section I present anecdotal evidence that the thought does occur in mundane settings, and I add examples of this and related kinds throughout the remainder of the book.

That the thought, or something like it, does occur in mundane settings is not important as proof of the mundanity of the thought. The thought could assert a mundane statement while still requiring as inspiration an awe-inspiring setting. Rather, insofar as the thought, or an experience like it, does occur in mundane settings, these occurrences allow us to eliminate one variable—awe at the object's intrinsic qualities—that may interfere with our ability to perceive other germane features of the thought.

I next describe three examples of my own of what appears to be Freud's thought in a mundane setting and one additional class of examples that highlights another possible kind of setting that may include either grand or mundane objects. A brief discussion follows.

## 2.1. The Examples

(1) One evening I attended a local town meeting concerning repairs that were to occur along a major road that runs near my house. Despite the inconvenience of the meeting and the anticipated nuisance of the repairs, both township officials and my neighbors showed a great deal of humor at the meeting. The scheduling of the repairs, moreover, was arranged to the residents' advantage, partly as a result of our input. For instance, we were to have privileged access to the road at some times of the day.

The next morning, customarily aggravated by the inevitable wait to turn onto the road in question during rush hour, I fondly recalled my neighbors' humorous allusions to this annoyance at the meeting; one neighbor had soliloquized, for instance, about the two-mile detour that he (allegedly) takes to avoid this turn. Turning and then proceeding along the road, I crossed an intersection that had been discussed in passing at the meeting. It has a hedge that blocks visibility and a traffic-flow problem caused by left-turning cars. Based upon an official assessment, the township engineer had reported that ample space existed for a left-turning lane and that the township would create one.

I cheerily noted all of these things as I passed by. "There's the hedge. Yes, there's room for a left-turning lane, etc." Once past the intersection, I had an obscure feeling of pleasure that, I felt, I might have expressed as "So it really *does* exist, the hedge!" or "The street *is* wider there!"

(2) In January of 1991, shortly before the U.S. bombing of Iraq in the Gulf War, I studied an atlas so that I could think more concretely about the physical location of the protagonists in the conflict (Iraq and Kuwait). My attention was eventually drawn to the map of Israel, which was on the same page. I noticed a small body of water called Yam Kinneret, also labeled "Sea of Galilee." I thought, "So *that's* the Sea of Galilee." Then, gazing away from the atlas, I had the thought, "So it really *does* exist."

These thoughts alluded to my days as a child studying Jewish history in a secular Jewish Sunday school. The force of the second thought ("Freud's thought") was that the Sea of Galilee was not just a biblical myth. It is obscure to me now whether I thought back then that the Sea of Galilee is real, or if it had been real in ancient times, whether it still exists today.

(3) While listening one morning to the radio station WQXR, I hear as I have before their weekly spot, "Inside *The New York Times* Best-Seller List." The announcer briefly synopsizes the first few items on the different

lists of the *Book Review* section of the Sunday *Times*. My interest is aroused by one of the titles. Later, while glancing through the *Book Review*, I turn to the best-seller list. There is the title. For a moment, I find myself surprised to find it there. I am unaware of having doubted or distrusted the announcer. I did not formulate any plan upon hearing the announcer to inspect the best-seller list when I picked up my newspaper. I just find it strange and delightful, for a moment, to discover the title on the list.

(4) I add to this series a class of examples that is already represented in part by the second of the preceding instances. One is prompted to have the thought on the basis of having seen not the object in itself, but other evidence of its existence. In the second of the preceding examples, I had (something that sounds like) the thought when I saw Yam Kinneret designated on the *map*. Although I do not recall having had the following experience precisely, it strikes me as potentially conducive to Freud's thought:

You walk past a landmark, for instance the remains of a Native American village, while on a hiking trip. Suppose that the site is labeled in some way such that there is no mistaking what it is. Now suppose that you later see the village designated on a map or pictured on a postcard. "So it *does* exist!" you might affirm, even though you already *saw* the site.

In a somewhat related vein, a few summers ago, impressed by appealing photographs that we had seen in a guidebook, my husband and I decided to take a hiking holiday in the Marble Mountains of far northern California (I mentioned this example in Chapter 1). Shortly before we left for our trip, some friends showed us pictures that they had taken of the area 10 years earlier when they had visited it. Moved by the close similarity between their pictures and the ones in the guidebook, I felt the thought slowly materialize that the mountains were indeed *real*. The thought was accompanied by the impression that I had not appreciated this fact earlier. Similarly, when I later attained my first view of the eponymous peak of the district from one of the trails, I was specially moved to see the thing of which I had seen pictures, in this case particularly appealing ones, including the picture on the guidebook's cover. The thought again materialized, "It's real."

Relevant for the class of instances that I am presently considering is the first of these two occurrences. Not the object (i.e., the district) in itself, but a set of pictures, generated independently of the first set of pictures that I saw, led me to have the "thought."

In between my perusal of the two sets of pictures, I had engaged in numerous activities that presupposed the existence of the area. I booked plane tickets and lodgings. I spoke at length to district rangers about trails and scenery. These behaviors both presupposed a firm conviction that the place exists and produced results—for instance, the rangers' advice and all of the pamphlets and maps that they sent me—that should have convinced me of the area's existence if I had felt at all skeptical. I had the "thought" subsequently, nonetheless.

## 2.2. Discussion

Each of the four examples, or groups of examples, that I have presented involves the implicit or explicit thought that such and such really exists, and in each but the second (Sea of Galilee, to be discussed presently), the observer clearly had, or in the case of the hypothetical examples, would have had, previous knowledge of the existence of the object. At least for the examples drawn from my own experience, the thought arose suddenly, with no clear antecedent, as did Freud's (own) thought.

Let us consider as canonical cases of the thought those occurrences of it that respond to one's firsthand contact with a grand and generally remote object, such as the Acropolis, of which one previously knew only indirectly. The examples that I described differ from canonical cases of the thought in the following ways:

The first and third examples (hedge/intersection and best-seller list) do not involve remote places, historical events, or any other reality resembling a "miracle" in any way. The objects involved are ordinary and close to hand.

In the second and fourth examples or sets of examples (Sea of Galilee, and map and picture examples), the thought is provoked by a token of the object (map or picture) rather than by direct, perceptual contact with the object. In the case of the hypothetical trek past the Native American village (fourth group of examples), perceptual contact occurred first.

Although it is not difficult to believe that the objects of the various examples exist, one *could* doubt their existence more easily than one would doubt the existence of the Acropolis, an object commonly known to exist. The hedge might not have existed, and the engineer could have erred in his assessment of the feasibility of a left-turning lane (first example). There is a good chance that I did not know the truth about the Sea of Galilee as a child (second example). As an adult, I also had not had a firm view; I had not thought about it. The radio announcer could have misreported the items on the best-seller list (third example). Although it may be difficult to dispute that a physical site is real once one has seen it (fourth example, the Native American village), prior to encountering it one may not have known that it existed.

Were these examples instances of Freud's thought, we could infer the following:

(1) The thought does not require as an object a remote place, historical event, or otherwise remarkable entity or event.

(2) Direct contact with the real thing may not be critical to the instigation of the thought. Hence, assuming that some uniformity exists across

cases, the thought is unlikely to express the idea that seeing something is special or that now that one sees it in its concrete tangibility, it superexists. What may matter is seeing an alternative, independently derived representation of it that would furnish evidence of its real existence, *were* there any doubt. Official records, especially maps, may provide one particularly evocative source of such representations.

(3) That the existence of the objects of any of the four thoughts (or groups of thoughts) that I have described could have been (or could be) doubted might suggest that the thought could, or does, presuppose doubt. However, the examples do not support this view. Although the information about the intersection could have been incorrect, I did not doubt it. I had seen the hedge before. As I drove through the intersection I noted all of the objects first, as if I fully expected to see them (e.g., "There's the hedge, there's the place for the left-turning lane . . .") and was glad only to be seeing mementos of the previous evening's unexpected revelry and success. *Then*, after a pause, I found myself exulting in the astounding discovery that these things really do exist. Although I could have opened the *Book Review* wondering whether I would see on the best-seller list the title that I had noted while listening to the radio, I did not wonder. Although one may feel that a place that one has just discovered (the Native American village, for instance) might not have existed for all that one knew, one now knows that it does. One has seen it.

In the case of the Sea of Galilee, it is possible that I was only correcting a piece of incomplete knowledge, as I did once when, after I had referred to Mount Olympus as mythical, a colleague informed me that it was real as well as fabled. However, my "thought" about the Sea of Galilee had the subjective trappings of the other examples. Although I did not know for sure that the Sea of Galilee existed, I am aware that at the moment that I had the thought (that the Sea of Galilee really does exist), I was not thinking about any former doubt or lack of knowledge. I had just observed the sea on the map, as though I expected it to be somewhere, only I did not know exactly where: "That's the Sea of Galilee." The reflection that it really does exist arose as an afterthought, as did the thought about the intersection, for instance.

With the possible exception of this ambiguous case, the series of examples would be no more easily explained by the hypotheses that I have previously rejected than would the canonical thought. "Common-sense" (face-value) explanations (Chapters 3 and 4) would still fail to account for the surprised acknowledgment of the existence of objects whose existence was not in doubt. Freud's account (Chapters 2 and 5) would again lack motivation: To drive through an intersection hardly suggests an accomplishment about which to feel victorious, guilty, or otherwise self-protective. To raise the alarmed "What if it didn't exist?" (Chapter 6) would make no sense in these cases. The consequences of these objects' not existing could hardly be deleterious. The existence of these objects, finally, far from miraculous or spectacular (this chapter), is unremarkable. Some, as a rule, would scarcely be noticed at all.

Whether or not they embody exactly the experience that one has when one has the thought in response to one's contact with awe-inspiring objects, these examples raise the same explanatory problems as do instances involving awe. Henceforth, therefore, in the interest of broadening the database for navigating these puzzles, I shall consider these mundane examples to be cases of the "thought."

# Chapter Eight

# *The Thought as Childlike: The Reencounter with the Known in a New Guise*

Whether it accompanies one's encounter with something truly grand or something commonplace, Freud's thought exults in a plainly obvious fact. It is mundane. One might, therefore, describe it as childlike: like something that a child would think or say. Children exult in what is obvious to adults, for it is not as obvious to them.

Scant though this redescription may seem, it could supply the missing "context" of the thought. Insofar as someone affirms what appears to be obvious because it is not, in fact, obvious to him or her, the context for an affirmation exists. It is *not* obvious that "all this really exists." The account that I shall develop in this chapter, therefore, is that the thought is childlike. When one has it, one adopts momentarily a child's point of view.

I distinguish this account from the proposal that I rejected in Chapter 3, section 1.3, that when they have the thought, people (adults) are replying to uncertainties that they may have had as children about whether the object of the thought exists. The latter proposal contained the incongruity that adults do not have these uncertainties. The current account holds that the adult speaks as a child, not as an adult. Therefore, the adult's knowledge is not relevant.

Whether this characterization of the thought as childlike does supply a possible context for the thought depends upon our establishing two things: that the thought reflects the vantage point of a child and that adults would return to that vantage point on occasions when they have the thought. I address these questions in turn.

## 1. The Childlike Vantage Point of the Thought

Children question or, as the case may be, affirm the obvious.

A tour group visiting the South Street Seaport in New York City first looked at a ship from the pier and then was taken on board. When the group reached the deck of the ship, a child asked the guide, "Are we there?" Evidently she was not sure whether the object upon which she stood was the same thing that she had seen from the pier.[1] No adult would wonder. This kind of error is common among children. The parameters within which they recognize the same object or scene are narrower than are the parameters within which adults do so.[2]

Taking his first plane ride, a different child, 4 years old, said to his father with combined relief and puzzlement, "Things don't really get smaller up here." He had seen planes only from the ground before.[3]

We might similarly imagine a *child* being taken on her first transatlantic plane ride after having learned about the various oceans and continents in school. Imagine *her* glee in seeing firsthand that "body" of water that "stretches" from North American to Europe.

Freud's thought consists, however, of a specific response to a specific circumstance. Insofar as the thought is childlike, adults return to this particular source of childlike interest and not to just any source. The examples of Chapter 7 lend insight into this source. I call it the "reencounter with the known in a new guise." The thought always occurs in this setting. One hears about the Acropolis in school and then sees it one day. One participates in a meeting in which a particular hedge is mentioned and then sees the hedge the next day. One sees the remains of a Native American village while out hiking and then sees the village marked on a map some time later. One sees one set of pictures of a potential vacation spot and then sees another set.

Normal adults usually find the reencounter with the same thing, in the same or a different guise, routine and unremarkable. We read the work of a colleague and then meet the person. We hear of a sale on garden hoses and then go buy one. Sometimes, however, even as adults we find reencounters with things or people that (or whom) we have known before exhilarating as well as strange and disconcerting. We *could* find it exciting and strange to meet the very colleague, Jill P., whose work we had read. As we approached the hoses on the shelf, we could suddenly find it exhilarating and strange actually to see and to possess one of the hoses that we saw in the ad. At such times we might suddenly think that Jill P. actually is a real person, or that the hoses mentioned—even depicted—in the ad really do exist.

Meeting the same thing after any kind of lapse or change can excite and intrigue us in this way. Having hiked away from a country lake where we are staying, we may find the view of it from the ridge that we have reached particularly gratifying, even if we fully expected to see it. Although returning to our house or garage normally does not arouse us, it could, especially after a lapse such as a vacation.

This feeling of pleasure in reencountering things seems to depend upon our having lost contact with the object (or event) at least briefly. The effect of sighting our lake again is diminished if, in the process of climbing up to the ridge, we never lose sight of the lake. We may still find it exciting to watch the lake diminish in size and to be able gradually to see its entire circumference in one glance. But we do not experience the feeling of something special and beyond the normal unless we have left the lake for a while. The experience is all the richer if we have not even thought about the lake during this intervening period.

When people have Freud's thought, they experience the object of the thought, lose it or turn from it, and then regain it. They lose it or turn from it in at least the sense that they do not think about it during some interval after their first encounter with it. Freud learned about the Acropolis years before he saw it. Then he saw it. I heard about the hedge one night and thought no more about it. I was struck by it (I had "Freud's thought") when I drove past it the next day. I heard the radio show about the items on the best-seller list at 9 o'clock one morning and forgot about it. I had a moment of realization—I had Freud's thought—when I encountered the actual list in the newspaper a few hours later.

Whereas adults take note of this experience of reencounter only infrequently, children often revel in the sort of repetition that I have been describing: seeing in real life a thing that has been talked about, discovering on a map or in a picture a place or an object that they have seen. Freud himself describes an instance of the complementary case, in which his 5-year-old son became upset when a family outing did *not* include the mountain (the Dachstein) that he had seen through a telescope and heard much discussed. When he had heard the mountain mentioned in connection with the planned walk (the walk was to depart from a town that lay at the foot of the Dachstein), Freud suspects, the boy

> had expected to climb the mountain ... and to find himself at close quarters with the hut *which there had been so much talk about* [italics added] in connection with the telescope. But when he found that he was being fobbed off with foothills and a waterfall, he felt disappointed and out of spirits.[4]

The pleasure in reencountering things dates from our infancy. Smiles register recognition of the reappearance of the same, important other people. Babies also delight in reproducing the same effect themselves.[5] Children can revisit the same spot, play the same game, or have the same story read to them endless times, with renewed, if not increasing, interest.[6]

Freud's thought exults, however, not in finding the same thing again, but in finding it in a new manifestation. This pastime also dates back, however. "Banana!" cries 1 1/2-year-old Rachel when she runs to the refrigerator to point to a banana after having seen a picture of one in a book.[7] Two-year-old Anne goes to retrieve her wooden hippopotamus when she and a parent read about a hippo in a book. Mention of a bicycle sometimes leads her to point to a window beneath which her father often keeps his bicycle.[8] In general, the matching of words to things preoccupies children in the second year,[9] as does the collecting in both talk and play of discrete exemplars of the same type of thing.[10]

Freud's thought conveys yet another sentiment, a feeling of one's self-importance: "*I* was there." Or "Where I was has official recognition." I have been to the Acropolis. I have seen, and indeed procured, one of the hoses that was described in the newspaper. The village that I walked past is on the map. Perhaps contributing to this sense of self-importance is an awareness of the public dimension of the image in question. The Acropolis is a well-known, public object, documented worldwide in texts, pictures, and films. A newspaper is an official, public organ, seen by innumerably many others. A map declares officially what is.

Rachel and Anne are too young to appreciate these wider entailments of the images that they encounter. They can take pride in showing an admiring parent what they know and might enjoy the approbation that results. Young children's collecting of their "images" may take place mostly in these dyadic settings.[11] This activity might nurture children's self-worth insofar as children receive, remember, or anticipate the resulting external approbation.

Before long, children appreciate the wider entailments that I have described. Freud's son wanted to climb the mountain that *everyone had talked of*. Another 5-year-old, Jakob, saw a building burn in Hamburg during a family outing there and then saw pictures of the blaze in the newspaper the next day. The pictures excited him dramatically and profoundly.[12] Mayor Rudolph Giuliani of New York City, a ubiquitous, public figure, shook the hand of 10-year-old Kyronne while a group from Kyronne's school visited City Hall. Kyronne proudly displayed his hand to his friend!(See Photos 8.1 and 8.2.)

## 2. Adult Contexts for the Return of the Childlike

Freud's thought contains ingredients that are recognizably childlike and therefore is appropriately characterized as childlike. These ingredients arise specifically in conjunction with the situation that elicits Freud's thought. The question remains of why adults would assume the point of view of a child in this situation.

PHOTO 8.1 **The Hand That Shook the Hand.** *In March of 1996, while stopping at the steps of City Hall to talk with schoolchildren, Mayor Rudolph Giuliani of New York City shook the hand of Kyronne Legette, 10. Photo by Ruby Washington. Reprinted by permission of the New York Times.*

PHOTO 8.2 *A delighted Kyronne showed his hand to a classmate. Photo by Ruby Washington. Reprinted by permission of the New York Times.*

Classically, psychoanalytic theory treats the return of the childlike in adults as irregular or pathological. Regressive, this return generally serves the purposes of defense. We have already rejected the idea that the thought, and hence by extension any childlike lapse that might support it, serves the purpose of defense.

Childhood tendencies may persist into adulthood, however, for the pleasure or other benefit that they bring, as Freud and others have recognized.[13] Why might this particular pleasure persist?

One's adoption of a child's naïveté and enthusiasm upon (re)encountering the expected might revitalize and refresh one's experience. Normally for us things dull from repetition. If, however, I can perceive the mountain that I stand on as maybe not the same thing that I saw through the telescope, or the telescope's image as maybe *un*connected with any real object, then I create for myself a surprise and interest that would otherwise be lacking.[14]

However, children find many things wondrous and surprising, and many childlike stances could rejuvenate our perspective. We could admire other simple aspects of the things we observe, for example, their shape or color. We could transform them imaginatively as children do and, for example, see the hut on the mountain as a gingerbread house or a shimmering flower as a fairy.[15] Why feel numinous wonder specifically at the real existence of things or at their reencounterability?

One can imagine why these would be central issues for *children*.

According to the tradition of so-called cognitive-developmental psychology initiated by the late Swiss psychologist Jean Piaget, establishing the "permanence" of objects (physical things and people) is a major occupation of the first years of life. *Permanence* refers to the idea that something that disappears can come back again. It still exists. Although the exact nature of what babies do and do not understand about the permanence of things is disputed by researchers, it is clear that babies' understanding in this area develops. Something—a bottle, a toy, a person—that a 2-year-old would consider retrievable when it disappears may be given up for lost by younger children when it disappears under the same circumstances.[16] Parallel developments occur throughout childhood.[17]

Both psychoanalytic and nonpsychoanalytic studies of early emotional development emphasize the profound impact upon babies' emotional life of the refinding of lost "objects."[18] *Object* here means love object, for example, mother or father. Babies must, according to Freud, "*rediscover*" in reality those objects that have given them satisfaction. They must "convince" themselves that these objects are "still there." A precondition of this effort is that objects that formerly afforded satisfaction "shall have been lost," at least for a while.[19] Games such as peek-a-boo and other hiding and finding games, each an incomparable delight to children, recreate this pattern of loss and recovery and hence enable mastery of the circumstance.[20]

To find, lose, recover, and to find again what is real as opposed only to remembered or imagined[21] are *the* central issues of early childhood. They elicit pain as well as pleasure. For some people, they are never fully resolved; possibly everyone harbors some lingering doubts. Perhaps when we have Freud's thought or otherwise exult in the (utterly expected) reencounter with something, we are responding to stored uncertainty from childhood about the recoverability of our earliest "objects."

One might wonder, though, why this stored uncertainty would prompt a specifically pleasurable response, namely Freud's thought, rather than, say, a neurotic response. A (relevant) neurotic response might consist, for example, in feeling anxious about the loss of insignificant objects, for instance, the bandanna that one packed for a walking tour.[22]

This ambiguity (as to the way in which this stored uncertainty might be played out) may mean only that the appropriate child analogue to Freud's thought is not the threat of loss of an object. It is the happy, playful retrieval of one, as occurs during games of peek-a-boo. Like the game of peek-a-boo, the thought expresses a reprieve from stored uncertainty over the fear of loss of objects. We exult that the object that we fully expected to meet again we did indeed recover.

This account makes the pathway to the thought somewhat long and indirect, although not inconceivable. A more direct, and to that extent more compelling, account would ascribe the thought to a more unequivocally positive impulse, or set of impulses, that survives more or less unchanged from our childhood. Evidence that relevant impulses survive is provided by adult activities, apart from those that resolve into Freud's thought, that reflect each of the three levels of "reencounter" experience that I distinguished for children in section 1 of this chapter.

At the first level, adults, as I noted, enjoy seeing the same thing again after a lapse: seeing the same lake again after one has hiked away from it, returning to one's garage after a vacation. Freud wrote that the potential for pleasure and comfort exists in the mere act of recognition, provided that recognition is not overly mechanized, as it is, for instance, in the act of dressing.[23]

At the second level, adults enjoy encountering diverse images of the same thing in activities such as birding or identifying wildflowers in the woods. Though they do not doubt that the birds that they have studied in pictures exist, bird-watchers find satisfaction in sighting a bird that they have seen only in their books. Conversely, people are enamored of replicas. Grownups as well as children enjoy model airplanes and model trains or would take pleasure in a small model of their home town, favorite city, or other locale that they had visited.

The role of the familiar—in a new guise—in our appreciation of things has been recognized at least since Aristotle, who explained the appeal of various art forms in this way. When we have known the object of a painting before we see the painting, we enjoy what Aristotle called the imitative aspect of the painting (*mimesis*) and not only its technical execution.[24]

The history of poetics continually revisits the theme that the artist performs precisely the mission of making the familiar strange again and hence of reinstating the "shock of recognition."[25] Invoking Aristotle's observation, Freud[26] suggests that jokes and related phenomena derive their appeal from the rediscovery of the familiar—a sound or idea—where something different might have been expected.[27] Nicholas Howe, commenting recently on the proliferation of travel literature (e.g., maps, pictures, films), remarks that the stereoscopic vision thereby afforded of a place embellishes our experience of it:

> This late in the day, no one can travel anywhere on the planet without having seen it before arriving. . . . To look both at the acquired images of a place and the place itself at the same time is a kind of stereopticon trick that adds the dimension of the past to the scene before one . . . .[28]

We also know the potential for the abuse of the pleasure in the familiar, whether in the same or a different guise, in the field of advertising, for example. We are naturally drawn toward the thing we know, sometimes against our better judgment.[29]

To the pleasure in recognition in itself, in the same or a different guise, may be added pleasure in reencounter at the third level that I distinguished earlier: the feeling of self-importance or self-participation that can attend one's encountering in the world or in the media an object that one has known, or seen, earlier. One might call this pleasure *benign narcissism*.

In our adulthood, no less than in our childhood, our participation in things makes them special. It drives what we notice in the world and what we do. I watch the three-meter diving finals in the summer Olympics one evening. The next morning I enjoy reading about this competition in the newspaper more than I do reading about the competitions that I did not see. To read about it is exciting in some way other than the way in which the event is exciting in itself (which it is not, especially, for me). As news, this item is the one in which I should have had the least interest. I already knew about it.

What is occurring here is clear. I enjoy reading about the event that I am familiar with, in which I was in some way personally involved.

The process works in reverse. The day after you read about a fire on the turnpike, you drive past the very location. It seems special, and you feel (in a way) pleased and proud, because you read about it.

In all of the instances that prompt Freud's thought, one encounters an object, or else evidence of an object, of which *one* knew earlier. The thought draws attention to this acquaintance in referring implicitly or explicitly to the object's existence "after all," that is, "just as *I* thought [learned, etc.]."

In recognizing this self-referring, even self-aggrandizing, aspect of the thought, I do not mean to denigrate it. If upon seeing the hose on the shelf, I can revel that it really exists after all, then I imbue the moment with more interest and significance than it would have otherwise. I relate it to my earlier experience, which in consequence also becomes more significant by virtue of its connection with the present.

One's experience of more significant objects also may become fuller and deeper insofar as one already knows them. Suppose, for instance, that, never having heard about it, you are shown the Tomb of the Unknown Soldier on a first trip to Washington, D.C. You might feel deeply moved by the sensitivity of those who conceived it and saddened by the horrors of war. Yet if you arrive having known about the edifice before, you may feel more deeply stirred, or stirred in a different way. This additional feeling has something to do with the object's already having had a location in your mind or with your already having had something to do with it.

The pleasure and satisfaction that we derive from the familiar and the urge to feel a part of things or feel our own importance never disappear. They need not be "stored up" to provoke a response during adulthood. Yet they are sufficiently primitive that they might prompt as unpremeditated and otherwise entrenched and obscure a response as Freud's thought. They account, moreover, for the thought's pleasurable affect. Unlike the accounts that might suggest a history of object loss or other uncertainty as a backdrop for the thought, this account invokes a positive impulse (self-feeling) rather than a negative one (fear).

This account not only is apparently able to explain the thought's paradoxical content and pleasurable affect but also avoids the contradictory expectations to which the aggregate of "common-sense" hypotheses led. I refer to the expectations, which I enumerated in Chapter 3, that the same assertion could mark the richness of the object relative to one's expectations of it or its correspondence with what one did anticipate. The thought does not concern these eventualities. It exults in finding the object again. This pleasure can arise regardless of whether the object is grand or mundane, and regardless of whether its qualities exceed, match, or disappoint one's expectations.

# Chapter Nine

# *Reencounter with the Known in Reality*

Freud's thought exults in one's finding the object again, not in just any incarnation but specifically in the real world or in some manifestation that connotes the object's real existence. "It is real" or "it exists," asserts the thought. We know why this discovery would interest children. Sometimes children are not quite sure of what exists in the real world and what does not, or of what will return in the real world and what will not. The discovery alone that words—and sometimes stories and pictures—correspond to something "out there" fascinates and excites.[1]

The question arises of why adults would find this discovery noteworthy. To be sure, on some occasions they need to "see to believe." Until you have your first child you may not fully "believe" that you could be a parent (or have a child). Until your colleague actually shows you the match between the suspect's fingerprints and the prints on the handgun, you may not fully "believe" that the match exists, given the weight that this evidence will carry. Some realities are too important or emotionally charged to preimagine or accept fully. Seeing helps us to believe.

Freud's thought occurs, however, when neither belief nor acceptance is in question. Despite the fact that one knows and fully accepts that the Acropolis exists, one might find oneself professing surprise at its existence when one saw it. Even if one did not profess this surprise (i.e., did not have Freud's thought), one would still find one's contact with the thing in itself (which I shall refer to also as "the real thing") special, again despite one's foreknowledge of the object's existence. The question arises not only of why the thought occurs, therefore, but also of why contact with the thing in itself would move someone who fully accepts the intelligence that the thing is real.

As I discussed in Chapter 1, contact with the real thing may on some occasions permit a sensory experience that one cannot attain otherwise.

Many people attest that one cannot fully know the sublimity, expanse, and beauty of the Grand Canyon unless one sees it. As may occur with the thought, however, people's awe of the real thing may be aroused when one's sensory experience of the object either has become degraded or has not changed. One would attach special importance to seeing the actual site of the Acropolis even if one saw it in fog, its structures covered with scaffolding, or its landscape teeming with people.[2] Alternatively, as I mentioned in Chapter 1, when a statue (of David) at the French Embassy was established almost certainly to be an authentic Michelangelo, New Yorkers gazed in a new way upon the self-same object that they had previously observed with less interest.

Imagine the converse to the latter situation. Thinking that you are looking at the real Mona Lisa, you feel all the awe that the experience engenders in so many people. An expert happens by and tells you that the piece is a copy. The awe disappears and may even transform to anger. As in the case of the Michelangelo statue of David, nothing in the *image* has changed.

People are moved in these various cases merely by the knowledge of whether the object in question is authentic (more generally, "real") and by whether *they* are seeing the authentic (viz., real) thing. To aid in the search for the thought's source, this chapter explores the question of why people are moved by these contacts.

## 1. Connecting with the Object's Connections

Walter Benjamin suggests one reason why people need to experience the authentic versions of things:

> The authenticity of a thing is the essence of all that is transmissible from its beginning, ranging from its substantive duration to its testimony to the history which it has experienced.[3]

To see the real thing is to come into contact with this causal chain.

People value this contact. Of people's attraction to the Michelangelo David, Barbara Grizzuti Harrison writes: ". . . fakes may look the same, but they are not the same. They have not been touched by the hand of the maker."[4] To touch the thing is to touch the hand. At an elegant auction house in New York, people purchased for vast sums not only fossils, but pieces composed of fossilized *droppings*. Said one participant, "The idea of being able to touch something that is 50 or 60 million years old just kind of gets to you."[5]

Sometimes one wants not only to touch but to reevoke the past with which an object or setting is associated:

People who visit Auschwitz often do so to connect with what occurred there. Neither skeptics nor those who would deny what happened there, they are believers. They visit the site neither to seek evidence of what occurred nor to clarify details; little evidence and few details remain.[6] Rather, to stand on the soil upon which the Jews trod and perished is to stand in a direct line with the events that occurred there. Traveling backward in time, *you* would encounter the Nazis, the gas chambers, and the suffering and horror of it all.

The same holds for positive realities. It is one thing to be reminded, by some contemporary event or passing thought, of the place where one grew up. It is another experience entirely to visit the playground, or even merely the location of the former playground, of one's youth. One feels something extraordinary, which as in visiting Auschwitz, stems not from the establishment of greater credulity as if via stronger proof, or from a clearer awareness of details, which may no longer exist. Indeed, if one has any difficulty recognizing the appropriate spot, or if, having recognized it, one wonders whether one really did frequent it, one will not feel deeply moved. But if one is sure, then one recreates another time and another self.

The distant "realities" that one seeks to evoke via the contact with contemporary objects can be fictitious, as I have already recounted. Some Japanese tourists travel to Prince Edward Island in Canada not only to visit the setting of Lucy Maude Montgomery's *Anne of Green Gables* series but even to be married in Montgomery's parlor, where she was married. "I've seen people get off the bus at Green Gables and cry," remarked one tourism agent.[7] Here the motive to confirm belief is out of the question. People want to connect with the story that they love, perhaps also with the hand that wrote it.

Nonetheless, people hanker to see the "real" thing when they have no special interest in the thing's history or other associations. Some who endure the crowds and the wait to see the Mona Lisa may not know who painted it or why it is a great painting, but they still want to see it and would be disappointed and irritated suddenly to be deprived of doing so. These people do not want to exist in a line with the events, or the hand, that produced the painting. They just want to see the real thing. Authenticity appears to matter in itself.

## 2. Connecting with the Personal Past

Authenticity matters more under some conditions than it does under others. Freud's son, whose disappointing journey I recounted in Chapter 8, wished to climb the Dachstein and see its hut, not to gain contact with the hand of any maker, but because he had seen these objects through his telescope. His own personal familiarity with these objects, *his* history with them, as opposed to *their* history, animated his desire to see them. Those who throng to see the Mona Lisa have seen it in *their* textbooks and guidebooks and have heard people talk about it. It lives in the public and the private mind.

As is shown in the following two examples, the real thing can move people, and move them deeply, when they have already been in some way immersed in it, when it has particular meaning for them.

In her eloquent book, *Longitude: The true story of a lone genius who solved the greatest scientific problem of his time*, Dava Sobel traces the story of John Harrison, the unlikely tradesman who first determined how to measure time precisely and continuously at sea and thus how to establish longitude. The face of the earth and of navigation were changed irrevocably as a result. With the ability to measure longitude, sailors could chart their location at sea regardless of whether they could see land. The scientific establishment of the day believed that longitude could be measured only relative to the stars. Harrison, therefore, faced the challenge not only of accomplishing an unprecedented mechanical feat, but of overcoming both the prejudice against him as an unknown and the resistance to the idea that his kind of solution held any promise at all.

It was only once her extensive research was well under way that Sobel, a U.S. citizen, traveled to England to see firsthand Harrison's clocks, now housed at the Maritime Museum at the Royal Observatory in Greenwich. Coming face to face with these machines at last, she writes

> —after having read countless accounts of their construction and trial, and after having seen every detail of their insides and outsides in still and moving pictures—reduced me to tears.[8]

Standing face to face with these objects, whose insides and outsides she knew in detail and whose story she knew so well, Sobel was *moved to tears*.

In a similar vein, recently two women who were subjects of two of the most famous photographs of the Vietnam War era attended as special guests a photography conference that memorialized photographers killed in the line of duty. One woman was Phan Thi Kim Phuc, who at the age of 9 was photographed at the center of a group of children running, burned and in agony, from a cloud of napalm. Soldiers followed, rifles in hand. The other was Mary Ann Vecchio, then a 14-year-old runaway, who was pictured, arms outstretched, next to a slain student during the confrontation between students and the National Guard at Kent State University, Ohio, in May of 1970. Readers may recall one or the other of these photographs. (See Photos 9.1 and 9.2.)

The appearance of the two women, more than 20 years later and hence grown and altered, brought to those at the workshop *"tears ... at the shock of recognition* [italics added], and at the slap of responsibility." Students at the meeting, who were too young to have lived through the events, were "still struck by meeting the people *from images they had grown up with* [italics added], and sharing the impact of the photographs. 'They were two of the most famous nonfamous people in the world,'" said [one of the students] . . . ."⁹ (See photos 9.3 and 9.4.)

To be sure, the two women could not have affected the audience as they did were it not for the historical events in which they were enmeshed and which they so poignantly evoked. But clearly operating also was the link between these women, as living realities, and the vivid images held in the past by both those who lived through the events and those who did not. Conceive in contrast a member of the audience who was learning for the first time of the Vietnam War and the era that surrounded it. Though saddened and dismayed by the atrocities, and perhaps admiring of the bravery of some, that person would seem unlikely to have been moved to tears.

Among those external realities that move us the most are the things with which we have a history, the things that are familiar to us and that evoke our former images, thoughts, expectations, and feelings. Conversely, we are moved when these images, thoughts, and so forth are given an external embodiment.

To be sure, other factors, such as firsthand exposure to the perceptual or historical "aura" (Benjamin's term) of a thing, may make contact with the real compelling on some occasions. Nonetheless, "reencounter" may operate in those cases also. People who stand in awe of the Grand Canyon, and who are intrigued especially to see what the "real one" looks like, have seen some version of the Grand Canyon before or heard about it. Although the New Yorkers who stand in awe of the Michelangelo statue of David may not have previous associations with that particular piece, they know who Michelangelo was. The people who arrive at Auschwitz and at Green Gables have detailed associations with these places and strong feelings about them. Although seeing my playground would mean a lot to me, it would probably mean little to you (except insofar as you have now "encountered" it in this book!) if you did not play in it.

## 3. Freud's Thought Revisited

It is the convergence between contact with an external object and one's past images of it that produces Freud's thought. Upon visiting the Acropolis, for example, you might feel strangely moved to tread in the very place where Socrates walked. But if you had the thought while there that "it really *does* exist . . . ," you would not be referring to this convergence of your and Socrates's perambulations. You would be alluding to your confrontation with the thing that occupied your schoolbooks or the thing of which your teachers spoke.

PHOTO 9.1  *South Vietnamese forces follow terrified children fleeing their village in South Vietnam after an accidental aerial napalm strike on the village in 1972. Kim Phuc, at center, had ripped off her burning clothes. Photo by Nick Ut. Reprinted by permission of AP/Worldwide Photos.*

PHOTO 9.2  *Mary Ann Vecchio's reaction at Kent State University in May of 1970. Photo by John Filo. Reprinted by permission.*

PHOTO 9.3  *Kim Phuc, the subject of a Pulitzer-prize winning photograph on the opposite page, placing flowers in 1995 on a memorial to photographers killed on the job. Photo by Gene Pierce. Reprinted by permission.*

PHOTO 9.4  *Mary Ann Vecchio, the subject of another Pulitzer-prize winning photo (opposite), at the same event. Photo by Scott Allen. Reprinted by permission.*

Although Freud's thought is more detached emotionally than is the more profound stirring of emotion evident in some of the preceding examples, it, too, might respond especially to some kind of intense preparation for the object.

Freud was classically educated and interested in antiquities as a hobby. Insights and observations of Greek dramatists and thinkers thread through his writings. He dedicated himself to the discovery of first principles in a way that mirrors the agenda of ancient thinkers. In standing on the Acropolis, he came face to face with all of *that*.

My mundane hedge (from the example that I first described in Chapter 7, section 2) evoked my memory of the good feelings and neighborly camaraderie at the meeting that I thought I would hate. *That* hedge then appeared in real life.

When I had the "thought" in response to seeing my friends' pictures of the Marble Mountains, I had been studying literature on the area and planning our trip, which by then was imminent.

But what of objects with which people have a more dispassionate and sketchy connection, for example, the hose lying in the garden shop, or the downed lamppost that you see after having read about its storm-induced fall in the newspaper? These examples lack the intensity of prior contact with the object that is evident in the foregoing examples. They retain only the correspondence between what one saw, learned, or knew before and what one encounters now as given in the real, outer world. We like this connection, as do children, and we appear to take note of it via the thought.

# Chapter Ten

# *The Adult Voice*

It may come as a surprise, therefore, that Freud's thought, although perhaps child*like*, is not something that a child would normally think, or say. Thus, the foregoing account of it requires modification. In this chapter, I attempt to resolve this new paradox, by trying to establish what the thought might mean such that an adult, but not a child, would express this meaning. I begin by documenting my claim that a child would not normally assert the thought.

## 1. Children and the Thought

Although the thought may correspond to one's recovery in the real world of a familiar object, the thought is no mere exultation in this recovery in itself. It extrapolates from this encounter that the object, therefore, really exists, and it expresses surprise at this finding. The following example highlights the difference between these assertions:

Suppose that, an occasional bird-watcher, you have just spotted your first wren outside your window. "It's a *wren!*" you cry, though you know that the species occurs with some regularity in the locality. Moments later, you reflect, "So they really *do* exist . . . ." The first of these utterances, "It's a *wren!*" exults in what I have called "reencounter." You are establishing that you know this bird, and you are excited to see a real specimen. The second comment, Freud's thought, reflects upon the first. It deduces a consequence from it: Wrens, therefore, really do exist.

Children assert the first of these comments ("It's a *wren!*") or its equivalent. Rachel's cry, at 1 1/2 years, of "Banana!" when she spotted the banana in the refrigerator is an instance. Children do not, however, make the latter extrapolation (Freud's thought), at least not in comparable circumstances, so far as I can tell from the available evidence.

Based upon a scan both of earlier diary studies and of modern databases of children's speech,[1] I know of no reported instances of Freud's thought or of any thought resembling it among children. Parents and teachers to whom I have described it do not recognize it as a child expression. Inherently it seems an uncharacteristic expression for a child, for two reasons:

First, children would seem unlikely to assume the reflective, backward-looking stance of the thought: "So [i.e., in the light of the evidence, and in contrast to what I might have thought earlier], all this really *does* exist. . . ." The allegation that a given proposition is "true after all" suggests a considered judgment, a balancing of different views, including a reappraisal of one's own former view. An extensive literature suggests that children come only gradually to juxtapose different vantage points spontaneously.[2]

Consistent with this pattern, children may ignore or forget their own previous point of view when the current evidence, or the opinion of others, contradicts it.[3] Insofar as children's past beliefs are so volatile, children would seem unlikely spontaneously to compare a current with a past impression, especially if they detect a possible inconsistency between them. Freud's thought contains exactly this comparison: "I see that it's real now, whereas perhaps I wasn't quite sure before." Even if the comparison is ultimately mistaken (the person had no doubt), children might not entertain even a mistaken thought of this form.

Given a new experience, children would seem unlikely to look back at all, whether correctly or incorrectly, to their former state of mind. They may recall some prior events when they are prompted, for example when an adult questions them.[4] They may reminisce spontaneously, for instance when going to sleep and engaging in "crib talk" and hence when *not* in the throes of a new experience.[5] Otherwise, children live in the present. Had Freud's son been taken up the Dachstein to the Simony Hütte rather than being "fobbed off" with foothills and a waterfall, he would not have mused, "So, it really *does* exist!" He would have explored the hut.

To be sure, under some conditions, children may exhibit some of the more adultlike tendencies to which I have alluded.[6] My point is that the thought's backward glance is not characteristically childlike.

Second, children would seem unlikely to contemplate the "existence" of something. Although even babies might be *described* as concerned with whether given objects exist, it is not clear that they code their concern in these terms. They may wonder whether the things or people that, or who, go away will come back, or they may become distressed in the things' or people's absence and joyous to see them again. But they could do all of these things without any reference to the more abstract problem of whether the things or people exist: have a real, external embodiment, are real as opposed to made up. Even once children are fully verbal, the root *exist* appears to be a late addition to their vocabulary.[7]

Children do, on the other hand, discuss the real existence of objects whose existence is seriously and pervasively in doubt. They argue over whether Santa exists (or is real). When the 4-year-old daughter of a colleague touched the late Princess Diana at a gathering, she was (jokingly) understood to be ascertaining whether Diana was "real."[8] A 7-year-old neighbor of mine insisted for months that he had been finding "real" Native American arrowheads in the woods nearby.[9] Though he meant real in the sense of authentic, as opposed to real in the sense of real-in-the-world rather than pictured in a book, he drew a pertinent contrast, made by children the same age and younger, with what is fake.[10]

Children probably would not, however, place the more ordinary things of their acquaintance in the same class as they would these controversial objects. A banana is not Santa. Even the 4-year-old who touched Princess Diana was understood, *jokingly*, by the *adults*, to be testing whether Diana was real. The child, at that age, was given to touching people whom she liked (see note 8).

By the age of 5 or 6, on the other hand, some children reflect philosophically upon the real existence even of everyday things. Wordsworth recalls having had to touch a tree or a wall when he was young to convince himself of its, and his, existence, although the exact age at which he did this is unclear.[11] "How do you know if something is true if you don't see it?" mused a different, philosophically minded 6-year-old. "Is it *real* if you can't see it? Like a cell. You can't see it with your eyes. Then is it *real*?" Some time later, upon glancing through a microscope at his first cell, he said, "Then is it more *real* when I see it now with the microscope than when I saw the onion skin just plain with my eyes?"[12] "How," asked another 6-year-old, "can we be sure that everything is not a dream?"[13]

As do discussions of Santa, these philosophical musings address consciously held questions. Freud's thought, which *sounds* as though it addresses a question, addresses none.

Not only different from the thought, these discussions and musings would seem unlikely to form the base from which the thought emerges later in adulthood. In the case of the discussions about controversial objects, one would be left to wonder why adults would suddenly elevate objects whose existence they do not doubt to the status not just of ordinary objects, but of *Santa*. As for the philosophical musings, the thought would seem unlikely to trace to them because they *are* consciously reflected musings, of the sort of which adults are also capable. Freud's thought is involuntary and unpremeditated, more like Rachel's gleeful cry of "Banana!" or Jakob's exultation in the Hamburg newspapers (which reported the fire that he saw) than like the considered "If I can't see it, then how do I know it's there?" But it *isn't* Rachel's or Jakob's responses.

Insofar as the thought occurs only to adults, one might wonder whether it merely camouflages a child's exultant glee in the recapture of the object (i.e., in "reencounter"): In our adult propriety, we would not exult "Banana!" or "It's the *Acropolis!*" Instead we demurely tip our hat and say that the thing exists after all.

This prospect is undermined by the circumstance that adults are capable of the more childish jubilation, as in the unreserved declaration "It's a *wren!*" Moreover, the thought may arise as a sequel to this more primitive assertion, as I noted at the beginning of this section. This circumstance suggests that the thought expresses a separate idea. It seems reasonable to infer, therefore, that the thought, no mere cover for the childish idea, expresses an experience that children do not have.

## 2. The Confirmation of Experience

The thought responds to some apparent uncertainty, generated on some basis, about the object's real existence. In comparable settings, children do not express this uncertainty. In what, then, might this uncertainty consist such that adults, but not children, acknowledge it?

We know, on the one hand, that the uncertainty cannot reflect any doubt that adults have now, or else have had recently, about the existence of the object. We eliminated that possibility by definition. Chapter 3 concluded that the uncertainty also could not be manufactured out of people's doubt about other matters, for example about the exact appearance and ambience of the object, or about the reliability in general of received information or one's own memory. The thought expresses surprise that *this* object exists, and it is the *object*, and not its derivative features, whose existence is surprising. We also eliminated the hypothesis (in Chapters 5 and 6) that the thought reflects neurotically generated uncertainty.

We decided, on the other hand, that the thought also does not respond to a *recalled* uncertainty about the object's reality, a doubt dating, for example, from one's childhood. In the case of many instances of the thought, such uncertainty never existed. It did not exist for Freud, regarding the Acropolis. It did not exist for me, regarding the hedge. Even if one did harbor such doubts earlier, the question would remain of why one would suddenly address this dated perception of the object rather than one's contemporary view. Chapters 8 and 9 offered the prospect that one simply becomes a child in the moment and, in that role, both raises and answers a child's question. We see now, however, that a child would not have the thought.

I propose that one *is* addressing one's earlier point of view; however, in juxtaposing this earlier point of view with one's current perspective, one distorts the earlier view. When they have the thought, adults, unlike children, revisit their earlier encounters with the object. When, upon reaching the top of the Acropolis, Freud declared that the site exists just as he learned at school, he recalled a previous (in fact the earliest) situation in which he knew the Acropolis. In noting suddenly that the hedge really does exist, when I drove past it, I recurred to the town meeting at which I had heard the hedge discussed. I want to suggest that the thought and its attendant feeling of unreality arise from this glance backward.

This account is suggested by reflection upon the related domain of experience of nostalgic reminiscence. Nostalgic images and feelings are elicited strongly by people's encounters with actual objects or places from their past.[14] For example, one visits after a long lapse the playground or neighborhood of one's youth or happens upon the ticket stub that one saved from a memorable movie that one saw long ago.

One may feel something extraordinary in these circumstances, as I noted in Chapter 9. It may seem odd to see the very playground, the very streets, or even merely the spot where the playground or the streets lay, now. One may feel surprise, consequently, that the streets (or other places) still exist.

Objectively, no grounds exist for this surprise. The location where the playground lay would not simply vanish; if the playground survived, *that* would afford no mystery either. If you stored the ticket stub in your high school yearbook and had not moved it since, its presence there would be unremarkable. Only your vantage point would make these conditions surprising.

That vantage point consists of the circumstance that you knew this same site or same object earlier, in a different time, and from a different vantage point (through the eyes of a 6-year-old, an adolescent, etc.). People cannot thoroughly revive this different time and perspective. Upon revisiting the river Wye, Wordsworth recalls (in his "Lines Composed . . . above Tintern Abbey . . . ") how hugely he felt in that location previously and how he could no longer feel exactly that way: "With many recognitions dim and faint,/And somewhat of a sad perplexity,/The picture of the mind revives again."[15] He could recall but not thoroughly reinhabit his former self. Thus (from "The Prelude"): ". . . the soul,/Remembering how she felt, but what she felt/Remembering not, retains an obscure sense/Of possible sublimity . . . ."[16]

One's failure on such occasions to accept entirely the reality of the standing objects that indeed have survived might reflect this "perplexity" over a lost time and lost self. During the earlier period and while inhabiting one's former self, one perceived these objects only as real. As one *looks back*, they seem not real.

Extending this scenario to Freud's thought, we might extrapolate: One's encounter, especially after a lapse, with a familiar object, which one knew previously in a different manifestation, evokes one's former experience of the object. Standing on the Acropolis, one recalls learning about the site, reading about it, perhaps thinking of it. Freud recalled being taught about it in school. One's revival of these experiences, now strongly demarcated as past or at least as "other" by the new and different manifestation of the ob-ject, creates a sense of unreality about the object. At the time when one first learned about it, one perceived it only as real. The *re*encounter with it makes it unreal.

In the case of Freud's thought, the feeling of unreality may be elicited not only, or not even, by a lapse in time between encounters with the object, but (also) by a shift in frame of reference after even a short time. We are startled, for example, if at a friend's party we meet the local librarian, whom we are accustomed to encountering at the checkout desk. I experienced the thought and its attendant "surprise" when I drove past the hedge but one day after I attended the meeting at which I had heard the hedge discussed.

When one has Freud's thought, however, one does more than experience a sense of unreality. One expresses a feeling of closure, a confirmation of something—ostensibly of the reality of the object. In a typical nostalgic episode, such as one elicited by the sight of one's old playground, one does not confirm anything, at least not explicitly. One remembers.

Yet nostalgic episodes may also confirm. One has recovered something that was lost—a part of self or perhaps time itself. In this recovery may lie a confirmation. One would be confirming, however, not that the standing remnants, of one's playground for example, really do exist, but that the pe-riod—the experiences—that they evoke really *did* exist.

People express affirmations of this kind when they have experienced gen-uine uncertainty about their past. Recently Werner Weiss, a computer-sys-tems consultant, created on the World Wide Web a virtual theme park, called "Yesterland," which presents attractions from Disneyland that no longer exist. Some visitors to the Web site report that they find themselves confirming their memories of attractions whose reality they had in fact come to question. Said one, "'Your Yesterland pages . . . cleared up one haunting memory. . . . I remember my 3-year-old brother being frightened by the drumming Indians. *I . . . had always thought I'd imagined them.* It was a *surprise* [italics added] to see the Indian Village on your pages.'"[17] Insofar as the past can become simply remote (as opposed to confused), as in the cases of nostalgic reminiscence that I have been discussing, contact with the past's relics may likewise serve to confirm the past.

Perhaps when one affirms the real existence of an object in cases of Freud's thought, one similarly confirms one's earlier experience. *That* real-ity we may be ready to question, though the question may not arise until our lives have progressed, however briefly or minimally, to a moment that demarcates our former experiences as separate or past.

Nonetheless, one affirms not that such and such an experience occurred but that its objects are real. One opposes "real" to one version or another of illusory and uses one's observation of new and independent evidence of the object to refute this possibility of illusion. If originally one encountered the object in a book, then when one has the thought, the object becomes not *only* an image in a book, but a thing whose reality is now firmly established by one's having seen it live, for example in a second book or in a film. If originally one saw the object (say, the Native American village past which one trekked) live, then it becomes not only something that one saw, but an entity whose reality is now established, for example, by a notation on a map or a listing in a guidebook.

All three of these features of the thought—its emphasis on the object, its refutation of one's sense of the object as illusory, and its glorification of independent evidence—seem to confirm the correspondence between one's experience and the external world. The question remains of why one would feel moved to confirm this correspondence when one had no fears of any disjunction. The impulse to confirm past experience seems easier to understand. One forgets, memory fades; people, and times, change.

Perhaps our way of validating our experience is simply to validate its correspondence with reality. We remain alert to signs of this correspondence not because we fear that we live solipsistically or have any other mistrust, but because we appreciate the feeling of connection. It gives meaning and purpose to our lives.

Thus, insofar as Freud's thought serves the purpose of confirming people's experience, the confirmation takes the form of affirming not only that the experience occurred, but that it was tied to reality. Like the feeling that one's experience may never have occurred, the feeling that the objects of one's experience were heretofore illusory is created on the spot. The possibility that the object might have attached only to oneself, one's perspective, or one's fantasies arises when one looks upon the scene from a second perspective. In that same glance, however, lies proof of the object's reality.

Not inclined to scan their experience backward and forward, children would not have this experience of the reality, or of the possible unreality, of a thing. They would experience the thing and perhaps enjoy the fact that they have it or mourn its loss when they do not have it. Consequently, they would not confirm the web of their experience. They would have no call to do so.

But it is paradoxical that based upon the same observation, we adults would recognize simultaneously that we had been living under an illusion (that the object is real) *and* that the illusory object is real after all. It would make more sense for one of these cognitions to follow the other. One recognizes, on some basis, that one has been living under an illusion. Then one recognizes, presumably on some new basis, that the object is real after all.

The thought effects this sequencing. One does not experience the separate realization "I was living under an illusion before!" One experiences a memory of this illusion. But the paradox remains: One feels surprise at the instantiation of a reality that one never thought of as doubtful or illusory.

Although they overlap to a noteworthy degree, therefore, nostalgic experiences and experiences of Freud's thought ultimately diverge. Standing nostalgically, after a lapse of a few decades, at the site of your old playground, you feel the past return partly, but not entirely. There may be confirmation and gratification in this experience, incomplete though it is. But you remain awkwardly suspended between past and present, which remain disjoint. You cannot go back, and yet for a moment you cannot assimilate the present fully either.

Freud's thought resolves. It effaces the incongruity between past and present (between one frame and another frame of experience). The old object and this new one are one and the same. That me and this me are the same. And, I connect with reality.

It is ironic that the experience that resolves its incongruity manifests paradoxically, whereas the one that does not resolve contains no paradox, only a lack of resolution.

# Chapter Eleven

---

# *The Paradoxical Stance*

While standing in full awareness that an object is real and never having suspected otherwise, normal adults can find themselves surprised to discover confirmation of the object's reality. Not only the thought's surprise, but the stance that people take in professing the surprise, is paradoxical. On the one hand, the stance is not genuine, for people do believe that the object exists. At the same time, the stance is not make-believe or otherwise deceptive. People are not acting a role as they would in a play or trying to trick either themselves or anyone else.

A child would not feel surprise at the discovery of evidence that something is real if the child knew already that the thing is real. At most, a child might pretend to be surprised, for example, in making a joke on a younger child or in producing a ruse to please or to manipulate a parent.

Nor is the thought's stance simply conflictual. When we are conflicted, we believe, feel, or have separate urges to do two incompatible things. For instance, we feel surprised for one set of reasons and unsurprised for other reasons that a friend won her swimming meet. People who have Freud's thought are not surprised at the object's existence on the basis of one set of expectations and unsurprised on the basis of a contrary set of expectations. They had one set of expectations, which included the understanding that the object is real. They then find themselves surprised to discover that it is real.

It is true that at the moment at which people find themselves surprised to discover that the object is real, they are also unsurprised to make this discovery; they knew that the object exists. This joint occurrence of surprise and unsurprise regarding the same circumstance may seem to accord with the idea that conflict underlies the thought. However, this "unsurprise" does not figure in the thought. The *thought* expresses surprise, though perhaps with a different coloring from the coloring of the surprise that one feels when one had genuinely doubted a reality (see Chapter 7, section 1).

Conceive, in contrast, a case of self-deception, in which owing to the presence of two contradictory urges, a person suppresses one of them. Aware, on the one hand, that one has had too much to drink, one determines, on the other hand, that one is fit to drive home from the party, when clearly one is not. We call this a case of self-deception. The puzzle arises: How can the same self both believe something and conceal it from itself? The answer that is usually offered is that the self splits in some way. Thus, for example, our tipsy partygoer looks for positive signs that he is fit to drive and suppresses the negative evidence. Or, his inhibitions gone altogether, and feeling good as he does, he perceives no obstacle to driving.[1]

Relevant for purposes of this investigation is the common intuition, as well as scholarly view, that the phenomenon of self-deception is to be explained by recourse to a conflict between impulses (and consequent suppression of one of them). So are the phenomena of indecision and ambivalence, for instance, explained with reference to the conflict between two contradictory impulses or feelings. In the latter cases, however, people may remain aware of both poles of the conflict. Insofar as one identifies a conflict, these various phenomena—outright conflicted opinion, self-deception, indecision, and ambivalence—cease to afford mysteries for most intents and purposes.[2]

Freud tried to understand his version of the "thought" on the model of conflict. As surely as he felt overjoyed to see Athens at long last, according to his analysis, he felt fearful and ashamed to admit his arrival. He had impulses, therefore, both to acknowledge and to deny his arrival. His thought, which both admitted and denied (although in the form of a past doubt) the real existence of the Acropolis, could be seen as a "compromise" between these two impulses.

Freud applied one version or another of this model of conflict to the explication of virtually all of the apparently paradoxical and otherwise (apparently) senseless behaviors that he analyzed. The conflicts whose presence he inferred resulted, according to his analyses, in the repression or other disavowal of one, and sometimes both, poles of the conflict. As a consequence, behavior (e.g., thoughts, feelings) arose that both exhibited signs of incongruity if not outright paradox and betrayed few clues as to its underlying source.

The delusional jealousy of the patient of Freud's whom I described in Chapter 5 observed this pattern, according to Freud. The woman, according to Freud's analysis, felt attracted to her son-in-law and also abhorred herself for the feeling—all of this unconsciously. The result, when convenient external provocation arose, was a senseless jealousy of her husband's supposed affair with another woman. The jealousy was senseless superficially not only in its lack of relation to fact, but also as a defense on the patient's part: Why would a woman who is happily married (as the patient was) invent an ungrounded story that could only bring her pain?

According to Freud's analysis, this jealousy settled the woman's conflict. On the one hand, it exonerated her for the attraction that she felt for her son-in-law. Her husband's alleged affair could legitimate her own unfaithful feeling. On the other hand, the jealousy allowed her to punish herself for her feeling, which she abhorred, through the pain that she felt at her husband's infidelity.

Freud thus gives a rational explanation of all of the constituents of the patient's superficially nonsensical symptom. It is clear not only from the conclusions of his individual analyses but also from his programmatic statements that the end of psychoanalytic inquiry in general was to produce a result of this kind: "The task is . . . to discover, in respect to a senseless idea and a pointless action, the past situation in which the idea was justified and the action served a purpose."[3]

Freud applied this conflict-repression model not only to pathological formations such as neurotic symptoms or other aberrant ones such as dreams, but also to apparently normal but somewhat obscure experiences and behaviors. Thus, he traced both the experience of the uncanny and the adult sense of the comic, which he believed (young) children could not experience, to the repression or other inhibition of childlike impulses.[4] This repression and other inhibition occurred naturally in the course of development.

Regarding the sense of the comic, for example, both an adult and a child might laugh at the raggedy tramp who stumbles and falls (an instance of slapstick). Whereas a child would laugh out of a sense simply of superiority, Freud thinks, adults' sense of the comic is obscure. We do not know why we laugh. This obscurity attaches, Freud theorizes, to our having surmounted and suppressed a child's reasons for laughing, while we still harbor those reasons just the same.[5]

In Chapter 5, I questioned whether, even when one could establish its presence, the type of psychic conflict that Freud believed underlay his thought could generate the thought. In many cases of the thought, however, one cannot identify a relevant conflict in response to which the thought might form. People have no reason to deny where they are and no reason to displace the recognition of their attainment onto the setting of the attainment.

The arguments in Chapters 8 through 10 would seem to rule out the applicability to the thought of Freud's model of the repression specifically of the childlike. To some extent, the thought just *is* childlike, at least superficially. It is naïve. To the extent that it is childlike, then it would not entail a *repression* of the childlike.

I examined (in Chapter 8) the possibility that the thought nonetheless might discharge people's stored (i.e., repressed) uncertainty from their childhood about the recoverability of their earliest (emotional) "objects." The connection seemed unhelpful at best, given the remoteness in time, subjective feeling, and content of early object-related anxieties from the thought.

Difficulties arose also with the hypothesis (which I considered in Chapter 10) that the thought camouflages the childlike delight in "reencounter": one's coming face to face with the familiar in a new setting. This hypothesis is undermined by the observation that adults express this more primitive pleasure directly.

The appropriateness of Freud's accounts of other phenomena aside, the model of repression—of an impulse that is held back or denied—does not appear to apply to Freud's thought. The thought does not appear to reduce to a conflict of any kind.

This inapplicability of the model of conflict to the thought has consequences for the nature of the thought. When a paradoxical or otherwise incongruous behavior, including a noticeably pathological one, can be traced to a conflict, the paradox or other incongruity disappears. One has simply two parties to a conflict. When no conflict can be identified, then the paradox (or other incongruity) remains.

I have argued not only that a relevant conflict cannot be identified for Freud's thought, but that the thought is inherently integrative and confirming. It confirms that all of the parts of oneself are there. It is the opposite of the expression of a divided self.

It is consistent with the preceding considerations that nostalgic reminiscence, but not experiences resulting in Freud's thought, would escape paradoxical expression. In the throes of a (nonpathological) nostalgic experience, one remains to some extent divided. One travels backward and to some extent remains oriented in the past. However, one remains partly oriented in the present, as a result of which the past, although more vivid than it had been previously, remains inaccessible. There is no paradox. There is only a tension that does not resolve.

In the absence of detailed analysis, I can point only tentatively to other experiences that may, as Freud's thought does, manifest paradoxically and coincide with the (momentary) integration, as opposed to the dissociation, of the self.

One tentatively related experience is tears of joy. Adults who receive a thoughtful although unexpected gift from a friend or spouse, or who at long last reach a heartfelt goal, may cry—for joy. Tears of joy are paradoxical. Normally one cries in sorrow or pain. Additionally, crying is generally associated with children, as is reflected, for example, in the expression "He cried like a baby." Like Freud's thought, however, crying for joy is not something that children ordinarily do. They cry in sadness or pain, or possibly relief.[6]

It would be easy to argue that tears of joy perform the function of acknowledging that a given situation is "too good to be true" (for purposes of this discussion, I am restricting the relevant cases to those in which the tears arise in appreciation of some welcome event that has befallen one, as opposed to arising in response to fictional situations such as the happy ending of movies and as opposed to crying as an expression of pure relief from effort or strain). One cries, according to this account, because one would not have expected to encounter, and more strongly feels that one ought not to have encountered, the situation that has arisen.

Perhaps tears arise on some happy occasions for reasons of this kind. But they also arise in situations that parallel the circumstance of Freud's thought. As I recounted in Chapter 9, Dava Sobel *cried* when she saw Harrison's clocks, as do some of the Japanese tourists who disembark at Prince Edward Island to see the setting of their beloved *Anne* stories. These tears are (or were) shed not in sadness or pain, but in gladness, the grief that Sobel might have felt for Harrison's struggles notwithstanding. Even the tears that were shed by participants at the photography conference at which the two "famous-nonfamous" photographic subjects appeared did not mourn the atrocities of which the two women had fallen victim. The tears arose, according to the journalist who observed them, ". . . at the shock of [the participants' own] recognition," as well as at the "slap of responsibility" that they felt.[7]

These tearful irruptions do not appear to express the sentiment that the events in question are "too good to be true." Nor could one easily identify the counterexpectation, or conflict, that might underlie this sentiment, insofar as people felt it. The sight of Harrison's clocks marked for Sobel the completion of a journey or at least the validation of a journey in progress. Although a Western cynic might not share the fascination of the Japanese for the *Anne* series, visiting her turf makes the enthusiasts feel whole for a moment. To experience the *shock* of recognition and the *slap* of responsibility that the women from the Vietnam-era pictures might evoke could be painful. However, both to recognize the same thing after more than two decades of changes and to accept responsibility for it (i.e., to make it one's own) affirm and complete both self and object.

Naomi Fein, the writer whose visit to the courtroom of the Scopes "Monkey Trial" I mentioned in Chapter 1, experienced an "excitement that *edged into anxiety*" as she neared the town. She felt "foolish" and "emotionally flat" as she approached the courtroom itself.[8] Rather than cry at an approaching moment of personal convergence, she felt dread.

Whether only via Freud's thought or in other ways as well, momentary epiphanies of integration—of self and self or of self and world—seem to assume a paradoxical form. Insofar as apparently paradoxical experiences can be traced to conflict, the paradoxes dissolve. If in a moment of integration we come together with ourselves, then perhaps the result must manifest paradoxically.

# Appendix: A Disturbance of Memory on the Acropolis

## An Open Letter to Romain Rolland on the Occasion of His Seventieth Birthday

TRANSLATED BY JAMES STRACHEY

My dear Friend,

I have been urgently pressed to make some written contribution to the celebration of your seventieth birthday and I have made long efforts to find something that might in any way be worthy of you and might give expression to my admiration for your love of the truth, for your courage in your beliefs and for your affection and good will towards humanity; or, again, something that might bear witness to my gratitude to you as a writer who has afforded me so many moments of exaltation and pleasure. But it was in vain. I am ten years older than you and my powers of production are at an end. All that I can find to offer you is the gift of an impoverished creature, who has 'seen better days'.

You know that the aim of my scientific work was to throw light upon unusual, abnormal or pathological manifestations of the mind—that is to say, to trace them back to the psychical forces operating behind them and to indicate the mechanisms at work. I began by attempting this upon myself and then went on to apply it to other people and finally, by a bold extension, to the human race as a whole. During the last few years, a phenomenon of this sort, which I myself had experienced a generation ago, in 1904, and which I had never understood, has kept on recurring to my mind.[1] I did not at first see why; but at last I determined to analyse the incident—and I now present you with the results of that enquiry. In the process, I shall have, of course, to ask you to give more attention to some events in my private life than they would otherwise deserve.

Every year, at that time, towards the end of August or the beginning of September, I used to set out with my younger brother on a holiday trip, which would last for some weeks and would take us to Rome or to some other region of Italy or to some part of the Mediterranean sea-board. My brother is ten years younger than I am, so he is the same age as you—a coincidence which has only now occurred to me. In that particular year my brother told me that his business affairs would not allow him to be away for long: a week would be the most that he could manage and we should have to shorten our trip. So we decided to travel by way of Trieste to the island of Corfu and there spend the few days of our holiday. At Trieste he called upon a business acquaintance who lived there, and I went with him. Our host enquired in a friendly way about our plans and, hearing that it was our intention to go to Corfu, advised us strongly against it: 'What makes you think of going there at this time of year? It would be too hot for you to do anything. You had far better go to Athens instead. The Lloyd boat sails this afternoon; it will give you three days there to see the town and will pick you up on its return voyage. That would be more agreeable and more worth while.'

As we walked away from this visit, we were both in remarkably depressed spirits. We discussed the plan that had been proposed, agreed that it was quite impracticable and saw nothing but difficulties in the way of carrying it out; we assumed, moreover, that we should not be allowed to land in Greece without passports. We spent the hours that elapsed before the Lloyd offices opened in wandering about the town in a discontented and irresolute frame of mind. But when the time came, we went up to the counter and booked our passages for Athens as though it were a matter of course, without bothering in the least about the supposed difficulties and indeed without having discussed with one another the reasons for our decision. Such behaviour, it must be confessed, was most strange. Later on we recognized that we had accepted the suggestion that we should go to Athens instead of Corfu instantly and most readily. But, if so, why had we spent the interval before the offices opened in such a gloomy state and foreseen nothing but obstacles and difficulties?

When, finally, on the afternoon after our arrival, I stood on the Acropolis and cast my eyes around upon the landscape, a surprising thought suddenly entered my mind: 'So all this really *does* exist, just as we learnt at school!' To describe the situation more accurately, the person who gave expression to the remark was divided, far more sharply than was usually noticeable, from another person who took cognizance of the remark; and both were astonished, though not by the same thing. The first behaved as though he were obliged, under the impact of an unequivocal observation, to believe in something the reality of which had hitherto seemed doubtful. If I may make a slight exaggeration, it was as if someone, walking beside Loch Ness, suddenly caught sight of the form of the famous Monster stranded upon the shore and found himself driven to the admission: 'So it really *does* exist—the sea-serpent we've never believed in!' The second person, on the other hand, was justifiably astonished, because he had been unaware that the real existence of Athens, the Acropolis, and the landscape around it had ever been objects of doubt. What he had been expecting was rather some expression of delight or admiration.

Now it would be easy to argue that this strange thought that occurred to me on the Acropolis only serves to emphasize the fact that seeing something with one's own eyes is after all quite a different thing from hearing or reading about it. But it would remain a very strange way of clothing an uninteresting commonplace. Or it would be possible to maintain that it was true that when I was a schoolboy I had *thought* I was convinced of the historical reality of the city of Athens and its history, but that the occurrence of this idea on the Acropolis had precisely shown that in my unconscious I had *not* believed in it, and that I was only now acquiring a conviction that 'reached down to the unconscious'. An explanation of this sort sounds very profound, but it is easier to assert than to prove; moreover, it is very much open to attack upon theoretical grounds. No. I believe that the two phenomena, the depression at Trieste and the idea on the Acropolis, were intimately connected. And the first of these is more easily intelligible and may help us towards an explanation of the second.

The experience at Trieste was, it will be noticed, also no more than an expression of incredulity: 'We're going to see Athens? Out of the question!—it will be far too difficult!' The accompanying depression corresponded to a regret that it *was* out of the question: it would have been so lovely. And now we know where we are. It is one of those cases of 'too good to be true'[2] that we come across so often. It is an example of the incredulity that arises so often when we are surprised by a piece of good news, when we hear we have won a prize, for instance, or drawn a winner, or when a girl learns that the man whom she has secretly loved has asked her parents for leave to pay his addresses to her.

When we have established the existence of a phenomenon, the next question is of course as to its cause. Incredulity of this kind is obviously an attempt to repudiate a piece of reality; but there is something strange about it. We should not be in the least astonished if an attempt of this kind were aimed at a piece of reality that threatened to bring unpleasure: the mechanism of our mind is, so to speak, planned to work along just such lines. But why should such incredulity arise in something which, on the contrary, promises to bring a high degree of pleasure? Truly paradoxical behaviour! But I recollect that on a previous occasion I dealt with the similar case of the people who, as I put it, are 'wrecked by success'.[3] As a rule people fall ill as a result of frustration, of the non-fulfilment of some vital necessity or desire. But with these people the opposite is the case; they fall ill, or even go entirely to pieces, because an overwhelmingly powerful wish of theirs has been fulfilled. But the contrast between the two situations is not so great as it seems at first. What happens in the paradoxical case is merely that the place of the external frustration is taken by an internal one. The sufferer does not permit himself happiness: the internal frustration commands him to cling to the external one. But why? Because—so runs the answer in a number of cases—one cannot expect Fate to grant one anything so good. In fact, another instance of 'too good to be true', the expression of a pessimism of which a large portion seems to find a home in many of us. In another set of cases, just as in those who are wrecked by success, we find a sense of guilt or inferiority, which can be translated: 'I'm not worthy of such happiness, I don't deserve it.' But these two motives are essentially the same, for one is only a projection of the other. For, as has long been known, the Fate which we expect to treat us so badly is a materialization of our conscience, of the severe super-ego within us, itself a residue of the punitive agency of our childhood.[4]

This, I think, explains our behaviour in Trieste. We could not believe that we were to be given the joy of seeing Athens. The fact that the piece of reality that we were trying to repudiate was to begin with only a *possibility* determined the character of our immediate reactions. But when we were standing on the Acropolis the possibility had become an actuality, and the same disbelief found a different but far clearer expression. In an undistorted form this should have been: 'I could really not have imagined it possible that I should ever be granted the sight of Athens with my own eyes—as is now indubitably the case!' When I recall the passionate desire to travel and see the world by which I was dominated at school and later, and how long it was before that desire began to find its fulfilment, I am not surprised at its after-effect on the Acropolis; I was then forty-eight years old. I did not ask my

younger brother whether he felt anything of the same sort. A certain amount of reserve surrounded the whole episode; and it was this which had already interfered
with our exchanging thoughts at Trieste.

If I have rightly guessed the meaning of the thought that came to me on the
Acropolis and if it did in fact express my joyful astonishment at finding myself at
that spot, the further question now arises why this meaning should have been sub-
jected in the thought itself to such a distorted and distorting disguise.

The essential subject-matter of the thought, to be sure, was retained even in the
distortion—that is, incredulity: 'By the evidence of my senses I am now standing on
the Acropolis, but I cannot believe it.' This incredulity, however, this doubt of a
piece of reality, was doubly displaced in its actual expression: first, it was shifted
back into the past, and secondly it was transposed from my relation to the Acropolis on to the very existence of the Acropolis. And so something occurred which was
equivalent to an assertion that at some time in the past I had doubted the real existence of the Acropolis—which, however, my memory rejected as being incorrect
and, indeed, impossible.

The two distortions involve two independent problems. We can attempt to pene-
trate deeper into the process of transformation. Without for the moment particular-
izing as to how I have arrived at the idea, I will start from the presumption that the
original factor must have been a sense of some feeling of the unbelievable and the
unreal in the situation at the moment. The situation included myself, the Acropolis
and my perception of it. I could not account for this doubt; I obviously could not attach the doubt to my sensory impressions of the Acropolis. But I remembered that
in the past I had had a doubt about something which had to do with this precise locality, and I thus found the means for shifting the doubt into the past. In the
process, however, the subject-matter of the doubt was changed. I did not simply recollect that in my early years I had doubted whether I myself would ever see the
Acropolis, but I asserted that at that time I had disbelieved in the reality of the
Acropolis itself. It is precisely this effect of the displacement that leads me to think
that the actual situation on the Acropolis contained an element of doubt of reality. I
have certainly not yet succeeded in making the process clear; so I will conclude by
saying briefly that the whole psychical situation, which seems so confused and is so
difficult to describe, can be satisfactorily cleared up by assuming that at the time I
had (or might have had) a momentary feeling: '*What I see here is not real.*' Such a
feeling is known as a 'feeling of derealization' ['*Entfremdungsgefühl*'].[5] I made an
attempt to ward that feeling off, and I succeeded, at the cost of making a false pro-
nouncement about the past.

These derealizations are remarkable phenomena which are still little understood.
They are spoken of as 'sensations', but they are obviously complicated processes, attached to particular mental contents and bound up with decisions made about those
contents. They arise very frequently in certain mental diseases, but they are not unknown among normal people, just as hallucinations occasionally occur in the
healthy. Nevertheless they are certainly failures in functioning and, like dreams,
which, in spite of their regular occurrence in healthy people, serve us as models of
psychological disorder, they are abnormal structures. These phenomena are to be
observed in two forms: the subject feels either that a piece of reality or that a piece
of his own self is strange to him. In the latter case we speak of 'depersonalizations';

derealizations and depersonalizations are intimately connected. There is another set of phenomena which may be regarded as their positive counterparts—what are known as '*fausse reconnaissance*', '*déjà vu*', '*déjà raconté*' etc.,[6] illusions in which we seek to accept something as belonging to our ego, just as in the derealizations we are anxoious to keep something out of us. A naïvely mystical and unpsychological attempt at explaining the phenomena of '*déjà vu*' endeavours to find evidence in it of a former existence of our mental self. Depersonalization leads us on to the extraordinary condition of '*double conscience*',[7] which is more correctly described as 'split personality'. But all of this is so obscure and has been so little mastered scientifically that I must refrain from talking about it any more to you.

It will be enough for my purposes if I return to two general characteristics of the phenomena of derealization. The first is that they all serve the purpose of defence; they aim at keeping something away from the ego, at disavowing it. Now, new elements, which may give occasion for defensive measures, approach the ego from two directions—from the real external world and from the internal world of thoughts and impulses that emerge in the ego. It is possible that this alternative coincides with the choice between derealizations proper and depersonalizations. There are an extraordinarily large number of methods (or mechanisms, as we say) used by our ego in the discharge of its defensive functions. An investigation is at this moment being carried on close at hand which is devoted to the study of these methods of defence: my daughter, the child analyst, is writing a book upon them.[8] The most primitive and thoroughgoing of these methods, 'repression', was the starting-point of the whole of our deeper understanding of psychopathology. Between repression and what may be termed the normal method of fending off what is distressing or unbearable, by means of recognizing it, considering it, making a judgement upon it and taking appropriate action about it, there lie a whole series of more or less clearly pathological methods of behaviour on the part of the ego. May I stop for a moment to remind you of a marginal case of this kind of defence? You remember the famous lament of the Spanish Moors '*Ay de mi Alhama*' ['Alas for my Alhama'], which tells how King Boabdil[9] received the news of the fall of his city of Alhama. He feels that this loss means the end of his rule. But he will not 'let it be true', he determines to treat the news as '*non arrivé*'.[10] The verse runs:

> '*Cartas le fueron venidas*
> *que Alhama era ganada:*
> *las cartas echo en el fuego,*
> *y al mensajero matara.*'[11]

It is easy to guess that a further determinant of this behaviour of the king was his need to combat a feeling of powerlessness. By burning the letters and having the messenger killed he was still trying to show his absolute power.

The second general characteristic of the derealizations—their dependence upon the past, upon the ego's store of memories and upon earlier distressing experiences which have since perhaps fallen victim to repression—is not accepted without dispute. But precisely my own experience on the Acropolis, which actually culminated in a disturbance of memory and a falsification of the past, helps us to demonstrate this connection. It is not true that in my schooldays I ever doubted the real existence of Athens. I only doubted whether I should ever see Athens. It seemed to me beyond the realms of possibility that I should travel so far—that I should 'go such a long way'. This was linked up with the limitations and poverty of our conditions of life in my youth. My longing to travel was no doubt also the expression of a wish to escape from that pressure, like the force which drives so many adolescent children to run away from home. I had long seen clearly that a great part of the pleasure of travel lies in the fulfilment of these early wishes—that it is rooted, that is, in dissatisfaction with home and family. When first one catches sight of the sea, crosses the ocean and experiences as realities cities and lands which for so long had been distant, unattainable things of desire—one feels oneself like a hero who has performed deeds of improbable greatness. I might that day on the Acropolis have said to my brother: 'Do you still remember how, when we were young, we used day after day to walk along the same streets on our way to school, and how every Sunday we used to go to the Prater or on some excursion we knew so well? And now, here we are in Athens, and standing on the Acropolis! We really *have* gone a long way!' So too, if I may compare such a small event with a greater one, Napoleon, during his coronation as Emperor in Notre Dame,[12] turned to one of his brothers—it must no doubt have been the eldest one, Joseph—and remarked: 'What would *Monsieur notre Père* have said to this, if he could have been here to-day?'

But here we come upon the solution of the little problem of why it was that already at Trieste we interfered with our enjoyment of the voyage to Athens. It must be that a sense of guilt was attached to the satisfaction in having gone such a long way: there was something about it that was wrong, that from earliest times had been forbidden. It was something to do with a child's criticism of his father, with the undervaluation which took the place of the overvaluation of earlier childhood. It seems as though the essence of success was to have got further than one's father, and as though to excel one's father was still something forbidden.

As an addition to this generally valid motive there was a special factor present in our particular case. The very theme of Athens and the Acropolis in itself contained evidence of the son's superiority. Our father had been in business, he had had no secondary education, and Athens could not have meant much to him. Thus what interfered with our enjoyment of the journey to Athens was a feeling of *filial piety*. And now you will no longer wonder that the recollection of this incident on the Acropolis should have troubled me so often since I myself have grown old and stand in need of forbearance and can travel no more.

I am ever sincerely yours,

SIGM. FREUD
January 1936

## Notes

Grateful acknowledgment is made to Sigmund Freud Copyrights, The Institute of Psycho-Analysis and The Hogarth Press for permission to quote from *The Standard Edition of the Complete Psychological Works of Sigmund Freud*, Vol. XXII (London: Hogarth, 1964), pp. 239–248, translated and edited by James Strachey. Originally printed in the U.S. by Basic Books, *Sigmund Freud: Collected Papers*, Vol. 5, J. Strachey (ed.), 1959, pp. 302–312.

1. [Freud had made a short allusion to the episode some ten years earlier, in Chapter V of *The Future of an Illusion* (1927c), but had not put forward the explanation.]

2. [In English in the original.]

3. [Section II of 'Some Character-Types Met with in Psycho-Analytic Work' (1916d).]

4. [Cf. Chapter VII of *Civilization and its Discontents* (1930a).]

5. [The word has been rendered variously into English. Henderson and Gillespie, *Text-Book of Psychiatry* (10th ed., 1969), use the term 'derealization', and make the same distinction as Freud between it and 'depersonalization' (Freud's *'Depersonalization'*).]

6. [Freud discussed these phenomena twice at some length: in Chapter XII (D) of *The Psychopathology of Everyday Life* (1901b), P.F.L., 5, 328–32, and in a paper on *'Fausse Reconnaissance'* (1914a). Cf. also the Wolf Man's 'veil' (1918b), P.F.L., 9, 311 and 339 ff.]

7. [The French term: 'dual consciousness'.]

8. [Anna Freud, *The Ego and the Mechanisms of Defence* (1936).]

9. [The last Moorish King of Granada at the end of the fifteenth century. Alhama, some twenty miles distant, was the key fortress to the capital.]

10. [Freud used the same phrase to describe the defensive process in Section 1 of his first paper on 'The Neuro-Psychoses of Defence' (1894a), and again in Chapter VI of *Inhibitions, Symptoms and Anxiety* (1926d), P.F.L., 10, 275 and *n.* 1.]

11. ['Letters had reached him telling that Alhama was taken. He threw the letters in the fire and killed the messenger.']

12. [The story is usually told of his assumption of the Iron Crown of Lombardy in Milan.]

## References

Freud, A. (1936). *The ego and the mechanisms of defence.* London, 1937; New York: International Universities Press, 1946.

Freud, S. (1894a). The neuro-psychoses of defence. *The Standard Edition of the Complete Psychological Works of Sigmund Freud, Vol. 3.* (J. Strachey, general Ed.; London: Hogarth, 1962, pp. 43–61).

Freud, S. (1901b). *The psychopathology of everyday life. Standard Edition, Vol. 6* (J. Strachey, general Ed.; London: Hogarth, 1960); Pelican Freud Library, Vol. 5 (A. Richards, Vol. Ed.; Harmonsworth, England: Penguin Books, 1985.).

Freud, S. (1914a). Fausse reconnaissance (*"déjà raconté"*) in psycho-analytic treatment. *Standard Edition, Vol. 13* (J. Strachey, general Ed.; London: Hogarth, 1955, pp. 201–207).

Freud, S. (1916d). Some character types met with in psycho-analytic work. *Standard Edition, Vol. 14* (J. Strachey, general Ed.; London: Hogarth, 1957, 309–333); Pelican Freud Library, Vol. 14 (A. Dickson, Vol. Ed.; Harmonsworth, England: Penguin Books, 1985), pp. 291–320.

Freud, S. (1918b) From the history of an infantile neurosis. *Standard Edition, Vol. 17* (J. Strachey, general Ed.; London: Hogarth, 1957, pp. 3–122); Pelican Freud

Library, Vol. 9. (Vol. Ed., A. Richards. Harmonsworth, England: Penguin Books, 1984), pp. 227–366.

Freud, S. (1926d). Inhibitions, symptoms, and anxiety, London, 1960. *Standard Edition, Vol. 20* (J. Strachey, general Ed.; London: Hogarth, 1959, pp. 77–174); Pelican Freud Library, Vol. 10 (Vol. Ed., A. Richards; Harmonsworth, England: Penguin Books, 1983), pp. 229–333.

Freud, S. (1927c). The future of an illusion. *Standard Edition, Vol. 21* (J. Strachey, general Ed.; London: Hogarth, 1961, pp. 3–56); *Pelican Freud Library, Vol. 12* (A. Dickson, Vol. Ed.; Harmonsworth, England: Penguin Books, 1985), pp. 179–242.

Freud, S. (1930a). Civilization and its discontents, New York and London, 1963; *Standard Edition, Vol. 21* (J. Strachey, general Ed.; London: Hogarth, 1961, pp. 59–157); *Pelican Freud Library, Vol. 12* (A. Dickson, Vol. Ed.; Harmonsworth, England: Penguin Books, 1985), pp. 243–340.

Henderson, D.K., and Gillespie, R.D. (1969). *A textbook of psychiatry* (10th ed.). London: Oxford University Press.

# Notes

## Introduction

1. J. Gorman, "Consciousness studies: From stream to flood," *The New York Times*, April 29, 1997, pp. C1, C5; see, for example, D. Chalmers, *Conscious mind: In search of a fundamental theory* (New York: Oxford University Press, 1996); D. Dennett, *Kinds of minds: Toward an understanding of consciousness* (New York: Basic Books, 1996).

2. E.g., R. Penrose, *The emperor's new mind* (New York: Oxford University Press, 1989); J. R. Searle, *The rediscovery of the mind* (Cambridge, MA: MIT Press, 1992).

3. See, for example, J. S. Bruner, *Acts of meaning* (Cambridge, MA: Harvard University Press, 1990), for relevant commentary.

4. S. Freud, "Instincts and their vicissitudes" (1915a), In S. Freud, *General psychological theory* (New York: Collier-Macmillan, 1963), p. 83.

5. S. Freud, "The 'uncanny'" (1919), *The psychopathology of everyday life* (1901), *Jokes and their relation to the unconscious* (1905), *Group psychology and the analysis of the ego* (1921), and *Introductory lectures on psychoanalysis*, Part III (1917), respectively.

6. S. Freud, *Group psychology and the analysis of the ego* (1921).

7. S. Freud, "The 'uncanny'" (1919).

8. S. Freud, *Jokes and their relation to the unconscious* (1905), *The psychopathology of everyday life* (1901), and *Introductory lectures on psychoanalysis*, Part III (1917) and Freud and Breuer, *Studies on hysteria* (1893–1895), respectively.

9. S. Freud (1917/1977), p. 25.

## Chapter One

1. S. Freud, "A disturbance of memory on the Acropolis: An open letter to Romain Rolland on the occasion of his seventieth birthday" (1936, reprinted in the appendix to this book, p. 80).

2. S. Freud, "Brief an Romain Rolland: Eine Erinnerungsstörung auf der Akropolis," *Gesammelte Werke*, 16 (London: Imago, 1950; original work published in 1936), p. 251.

3. S. Freud, "A disturbance of memory on the Acropolis" (1936, reprinted in the appendix to this book, p. 80).

4. S. Freud, "A disturbance of memory on the Acropolis" (1936, reprinted in the appendix to this book, pp. 79–86).

5. P. Lively, *Moon tiger* (New York: HarperCollins, 1989), p. 108.

6. See, for example, M. K. Johnson and C. L. Raye, "Reality monitoring," *Psychological Review*, 88 (1981, 67–85); E. F. Loftus, "The malleability of human memory," *American Scientist*, 67 (1979, 312–320).

7. See, for example, H. M. Sno and D. H. Linszen, "The *déjà vu* experience: Remembrance of things past?" *American Journal of Psychiatry*, 147 (1990, 1587–1595).

8. See note 6 of this chapter.

9. See, for example, S. Freud, *The psychopathology of everyday life* (1901), Chapter 12; Sno & Linszen, "The *déjà vu* experience . . . ." (1990).

10. Diagnosis 300.6 in *Diagnostic and statistical manual of mental disorders*, 4th ed. (*DSM-IV*; Washington, DC: American Psychiatric Association, 1994), pp. 488–490.

11. Also classified under diagnosis 300.6.

12. N. R. Kleinfield, "Death scene now is lure for curious," *The New York Times*, May 1, 1995, p. A1.

13. C. Trillin, "Anne of Red Hair: What do the Japanese see in Anne of Green Gables?" *The New Yorker*, August 5, 1996, pp. 56–61.

14. B. G. Harrison, "The master's hand," *The New York Times*, January 28, 1996, p. E13.

15. N. Fein, "At the true 'Trial of the Century,'" *The New York Times*, October 13, 1996, sec. 5, p. 25.

16. S. Freud, *The future of an illusion* (1927/1961), Chapter V, pp. 31–32.

17. S. Freud, *Introductory lectures on psychoanalysis* (1917), Chapter XVII.

18. Regarding the circumstances under which mental processes are unlikely to be accidental, Freud once wrote that he "found it impossible to believe that an idea produced by a [patient] while his attention was on stretch could be an arbitrary one . . . ." (*Five lectures on psychoanalysis*, New York: Norton, 1910/1961, p. 29.)

19. See, for example, S. Freud, "The unconscious" (1915b); F. Fromm-Reichmann, *Psychoanalysis and psychotherapy: Selected papers* (Chicago: University of Chicago, 1959), especially Part III on schizophrenia; R. D. Laing, *The divided self* (Baltimore: Penguin, 1965).

20. L. Wittgenstein, *Philosophical investigations*, 3rd. ed. (New York: Macmillan, 1958).

21. T. J. Luce, personal communication, March 1, 1993.

22. Discussions that dwell upon Freud's own experience include M. Ater, *The man Freud and monotheism* (Jerusalem: Magnes, 1992); J. F. Flannery, "Freud's Acropolis revisited," *International review of Psychoanalysis*, 7 (1980, 347–352); J. M. Masson and T. C. Masson, "Buried memories on the Acropolis: Freud's response to mysticism and anti-Semitism," *International Journal of Psychoanalysis*, 59 (1978, 199–208); C. Schorske, "Freud's Egyptian dig." *New York Review of Books*, XLI (10), May 27, 1993, 35–40; and H. Slochower, "Freud's *déjà vu* on the Acropolis: A symbolic relic of '*mater nuda*,'" *Psychoanalytic Quarterly*, 39 (1970, 90–102). R. I. Fried ("The Stendhal Syndrome: Hyperkulturemia," *Ohio Medicine*, 87, 1988, 519–520), I. B. Harrison ("A reconsideration of Freud's 'A disturbance of memory on the Acropolis' in relation to identity disturbance," *Journal of the American Psychoanalytic Association*, 14, 1966, 518–527), and N. Howe ("Reading

places," *Yale Review*, 18, 1993, 60–73) consider the general occurrence of a strange or special feeling in connection with one's seeing significant places that one knows about, though again not the occurrence of the "thought" in particular. I discuss Howe's remarks in passing in Chapter 8.

## Chapter Two

1. S. Freud, "A disturbance of memory on the Acropolis" (1936, reprinted in the appendix to this book, p. 80).

2. S. Freud, "Some character-types met with in psychoanalytic work," (1916), sec. II.

3. M. S. Horner, "Femininity and successful achievement: A basic inconsistency," in D. J. Bardwick, E. L. Douvan, M. S. Horner, and D. Gutmann (Eds.), *Feminine personality and conflict* (Pacific Grove, CA: Brooks/Cole, 1970); see also W. Swann, *Self traps: The elusive quest for higher self-esteem* (New York: Freeman, 1996).

4. S. Freud, "A disturbance of memory on the Acropolis" (1936, reprinted in the appendix to this book, p. 81).

5. One may wonder about the accuracy of Freud's recounting, especially of the subjective aspects of his experience on the Acropolis, given the lapse of 32 years in between the incident and his reporting of it. In 1909, however, 5 years after the trip to Athens, Freud alluded to a similar configuration of feelings regarding his overcoming, finally, of his reluctance to travel to Rome, which he had long wished to see: "I discovered . . . that it only needs a little courage to fulfill wishes which till then have been regarded as unattainable" (footnote 2, added in 1909 to S. Freud, *The interpretation of dreams* [1900/1965], Chapter 5, sec. B, p. 226). Whether or not Freud accurately attributes these feelings to his experience on the Acropolis in 1904, we at least have evidence that he experienced feelings like them on another occasion at around the same time.

## Chapter Three

1. B. Russell, "Knowledge by acquisition and knowledge by description," in *The problems of philosophy* (Buffalo, NY: Prometheus, 1988), pp. 46–59.

2. See, for example, D. C. Riccio, V. C. Rabinowitz, and S. Axelrod, "Memory: When less is more," *American Psychologist*, 49 (1994, 917–926).

3. J. L. Austin, *Sense and sensibilia* (London: Oxford University Press, 1962).

4. This musing is from the ancient Chinese philosopher Chuang-Tzŭ, *Inner chapters* (trans. G. Feng and J. English; New York: Vintage/Random House, 1974), Chapter 1, p. 5.

## Chapter Four

1. For discussion of this class of expressions, see, e.g., J. Searle, "Indirect speech acts," in P. Cole and J. L. Morgan (Eds.), *Syntax and semantics: Volume 3. Speech acts* (New York: Academic Press, 1975), pp. 59–82.

## Chapter Five

1. S. Freud, *The psychopathology of everyday life* (1901), Chapter 1.

2. S. Freud, *Introductory lectures on psychoanalysis* (1917), Lecture XVII.

3. S. Freud, *Introductory lectures on psychoanalysis* (1917).

4. Displacements are instances of what Freud calls "primary process" thought. Primary process thought is guided by the simple association between ideas without regard for logical coherence. It has no aim toward argument or explanation. It typifies unconscious mental functioning, according to Freud. For a discussion of primary process thought and its antithesis, secondary process thought, see, for example, S. Freud, *The interpretation of dreams* (1900), Chapter 7, sec. E, or S. Freud, *An outline of psychoanalysis* (1940), Chapters IV–V.

5. S. Freud, "A disturbance of memory on the Acropolis" (1936, reprinted in the appendix to this book, p. 82).

6. At the same time, Freud calls the feeling that he experienced a derealization, which, as a clinical condition, could be associated with severe anxiety (see *DSM-IV*, diagnosis 300.6, depersonalization disorder, pp. 488–490). It is conceivable, therefore, that he felt more disturbed by his supposed derealization than he conveys in his letter to Rolland.

If Freud did experience distress as a result of his derealization, then the course of my argument would change. I would be assessing whether the thought that ensued ("So all this really does exist . . .") was a plausible response to this fear, given Freud's assumptions about defense mechanisms (note the immediately following discussion in the text, which suggests that even if Freud did attempt to "displace" this feeling of unreality, other problems arise with his account of how this displacement might have occurred). Because Freud's paper does not depict any great disturbance, I shall not pursue this line of argument. Additionally, the experience that Freud describes, that of a mild feeling of unreality, is closer to the generic experience that people have on occasions of "Freud's thought" than is the feeling of fear and panic (see, however, R. I. Fried, "The Stendhal syndrome: Hyperkulturemia" [*Ohio Medicine, 87*, 1988, 519–520] for discussion of a more drastic line of reaction, a "history-depression" that overtakes some people when they visit famous, culturally saturated places).

7. S. Freud, *Introductory lectures on psychoanalysis* (1917/1977), Lecture XVI, pp. 248–254.

8. See, respectively, S. Freud, *Moses and monotheism* (1939) and S. Freud, "The unconscious" (1915b, reprinted in S. Freud, *General Psychological Theory*, P. Reiff, Ed., 1963), p. 149.

9. S. Freud, *Totem and taboo* (1913), Chapter III.

10. See my discussion of the first assumption of Freud's account, earlier in this section.

11. M. Freemantle, "Walking on air," *Chemical and Engineering News*, January 29, 1996, p. 34.

12. S. Freud, "A disturbance of memory on the Acropolis" (1936, reprinted in the appendix to this book, p. 82).

13. Researchers have identified a related strategy of "*defensive* pessimism" (italics added) on account of which people set unrealistically low expectations in risky

situations (e.g., dating, making their way through college; see J. Norem and N. Cantor, "Defensive pessimism: Harnessing anxiety as motivation," *Journal of Personality and Social Psychology, 51,* 1986, 1208–1217).

14. S. Freud, *The future of an illusion* (1927/1961), Chapter V; see the discussion of Freud's remark in Chapter 1 of this book.

## Chapter Six

1. My mention of an "aura" surrounding an authentic historical relic, such as the Acropolis, may evoke for some readers Walter Benjamin's account of the "aura" that surrounds authentic, as opposed to mechanically reproduced, objects of art (W. Benjamin, "The work of art in the age of mechanical reproduction," in *Illuminations* [Trans. H. Zohn], New York: Schocken Books, pp. 217–251). I discuss his conception in Chapter 9, sec. 1.

## Chapter Seven

1. S. Freud, "A disturbance of memory on the Acropolis" (1936, reprinted in the appendix to this book, p. 80).

2. S. Freud, "Brief an Romain Rolland: Eine Erinnerungesstörung auf der Akropolis," *Gesammelte Werke, 16* (London: Imago, 1950; original work published in 1936), pp. 251–252.

3. S. Freud, "A disturbance of memory on the Acropolis" (1936, reprinted in the appendix to this book, p. 80).

4. J. Darnton, "Fiction is stranger than truth," *The New York Times,* March 20, 1994, sec. 4, p. 1.

5. Goldwyn, S. (Producer), McLeod, N. Z. (Director) (1947), *The secret life of Walter Mitty* [film], (available from Home Box Office).

6. Along these lines, Hawaiians and arriving tourists were possessed by "ecliptomania" in advance of the 1991 total solar eclipse (T. Egan, "4 minutes of darkness puts a glow on Hawaii," *The New York Times,* July 3, 1991, sec. 1, p. 1). One does not hear tales of "Acroptomania."

7. We also may *not* wonder, in this case as in even some of the more extreme of the preceding examples. We might take Saturn's rings utterly for granted in the way that we do the Atlantic Ocean, for example (see the next paragraph of text). Insofar as we do not wonder, and find ourselves puzzling that "it really *does* have rings . . . !" we have had a bona fide occurrence of Freud's thought. I wish to suggest here that a response of this kind to a spectacle such as Saturn need not be a bona fide case of the "thought."

## Chapter Eight

1. E. Santner (personal communication, August 1992).

2. For demonstration of this limitation in babies, see, for example, P. L. Harris, "Infant cognition," in P. H. Mussen (Ed.), *Handbook of child psychology: Vol. 2. Infancy and developmental psychobiology* (4th ed., M. M. Haith and J. Campos,

Vol. Eds., 1983, pp. 689–782). For demonstration of residual errors in childhood, see M. Cole and E. Subbotsky, "The fate of stages past: Reflections on the heterogeneity of thinking from the perspective of cultural-historical psychology (paper presented at the symposium "The cultural environment in psychology," honoring Ernst Boesch. Merlingen, Switzerland, October 21–24, 1991); J. Piaget, *The construction of reality in the child* (New York: Basic Books, 1937), Conclusion; and J. Piaget and B. Inhelder, *The child's conception of space* (New York: Norton, 1948).

3. G. B. Matthews, *Philosophy and the young child* (Cambridge, MA: Harvard University Press, 1980), p. 2.

4. S. Freud, *The interpretation of dreams* (1900/1965), p. 161.

5. For relevant observations, see, for example, T. G. R. Bower, *Development in infancy* (2nd ed., San Francisco: Freeman, 1982), Chapter 8, and references cited there; J. Piaget, *The origins of intelligence in children* (New York: International Universities Press, 1936/1952); and especially J. Piaget, *The construction of reality in the child* (1937/1954), Chapter 3; also J. S. Watson, "Smiling, cooing, and the game" (*Merrill-Palmer Quarterly, 18,* 1973, 323–339). On repetition as a fundamental fact of psychological life, consider Freud, *An outline of psychoanalysis* (1940/1989), p. 17: "The state . . . which an organism has reached gives rise to the tendency to re-establish that state as soon as it is abandoned." Consider also Piaget's (*The origins of intelligence in children,* 1936/1952, p. 43) version: "The individual, on however high a level of behavior, tries to reproduce every experience he has lived."

6. "Children will never tire of asking an adult to repeat a game . . . till he is too exhausted to go on. And if a child has been told a nice story, he will insist on hearing it over and over again rather than a new one." S. Freud, *Beyond the pleasure principle* (1920/1989), p. 42.

7. I. B. Vogel, personal communication, January 1996.

8. R. Bittner, personal communication, August 1996.

9. See, for example, Roger Brown, *A first language: The early stages* (Cambridge, MA: Harvard University Press, 1973).

10. For example, S. Sugarman, *Children's early thought: Developments in classification* (New York: Cambridge University Press, 1983).

11. Research on the social context of children's language learning (e.g., J. S. Bruner, *Child's talk: Learning to use language,* New York: Norton, 1983; K. Nelson, *Making sense: The acquisition of shared meaning,* Orlando: Academic Press, 1985) describes the occurrence of exchanges evocative of these examples.

12. P. Gollwitzer, personal communication, May 1996.

13. Freud (*Jokes and their relation to the unconscious,* 1905) uses this conception to explain both jokes, which he construes as revivals of the word and idea play of childhood, and our enjoyment of fiction (S. Freud, *Creative writers and daydreaming,* 1908). Our enjoyment of fiction, he says, derives from our propensity to fantasize, itself an extension of our earlier impulse to play. S. J. Gould (*An urchin in the storm,* New York: Norton, 1987, p. 63) believes that the biological process of *neoteny,* the retention into adulthood of primitive structures through a slowing of developmental rate, extends to psychological properties, as in, according to Gould, the association between adult creativity and childhood wonder; see also H. Gardner (*Creating minds: An anatomy of creativity as seen through the lives of Freud, Ein-*

*stein, Picasso, Stravinsky, Eliot, Graham, and Gandhi,* New York: Basic Books, 1993).

14. Wordsworth's childhood "visions" were the key to his imaginative life as an adult. In his "Ode: Intimations of Immortality" (1807, in T. Hutchinson and E. de Selincourt, Eds., *Wordsworth: Poetical works,* New York: Oxford, 1990, 460–462), he portrays these visions as:

". . . obstinate questionings
Of sense and outward things,"
and as,
"Fallings from us, vanishings;
Blank misgivings of a creature
Moving about in worlds not realized . . . ." (lines 142–146)

15. See, for example, J. Sully, *Studies of childhood* (New York: Appleton, 1914), Chapter 3.

16. See, for example, J. Piaget (*The construction of reality in the child,* 1937), Chapter 1, and P. L. Harris, "Infant cognition," in P. Mussen, *Handbook of child psychology* (1983), pp. 689–782.

17. See the references in note 2 of this chapter.

18. For a review, see, for example, J. R. Greenberg and S. A. Mitchell, *Object relations in psychoanalytic theory* (Cambridge, MA: Harvard University Press, 1983).

19. S. Freud, "Negation," in *General psychological theory,* P. Reiff, compiler (New York: Norton, 1925), pp. 215–216.

20. See, for example, A. Freud, "About losing and being lost," *The psychoanalytic study of the child,* 22, (1967, 9–19); S. Freud, *Beyond the pleasure principle* (1920), Chapter II; J. R. Greenberg and S. A. Mitchell, *Object relations in psychoanalytic theory,* 1983).

21. S. Freud, "Formulations regarding two principles of mental functioning," in *General psychological theory,* P. Reiff, compiler (1911/1963), pp. 21–28.

22. See the discussion in the sources cited in note 21, this chapter, for elaboration.

23. S. Freud, *Jokes and their relation to the unconscious* (1905/1989), Chapter IV, p. 148; see also references cited there.

24. Aristotle, *Poetics,* 4 (*The basic works of Aristotle,* R. McKeon, Ed., New York: Random House, 1941).

25. See, for example, Pomorsky, *Language, poetry, and poetics: The generation of the 1890s: Jakobson, Trubetskoy, Majakovsky* (Berlin and New York: de Gruyter, 1987).

26. S. Freud, *Jokes and their relation to the unconscious* (1905), Chapter 4.

27. The most disarming jokes of this kind confront the listener with familiar material at every level. For instance, Garrison Keillor's farcical song "Oy, Chuck and Katie" (*Songs of the Cat,* BMG Music, 1992) will amuse any listener who knows about *cats.* It affects far more intensely any listener who recognizes the song as a parody of a melody of *Mozart's,* and who knows that the melody is the aria "*Voi che sapete,*" which rhymes with the title of the farcical song. Moreover, the farcical song is sung by *Frederica von Stade,* who is known for her rendition of the Mozart aria.

28. N. Howe, "Reading places," *Yale Review, 81* (1993), p. 68.

29. See, for example, P. Aitkin, "Television advertising and socialization to consumer roles" (Vol. 2) *U.S. Department of Health and Human Services* (1976); D. Cox and A. Cox, "What does familiarity breed?" *Journal of Consumer Research, 15* (1988, 111–116; R. Zajonc, "Attitudinal effects of mere exposure," *Journal of Personality and Social Psychology, Monograph Supplement*, Part 2, 9, No. 2 (1968, 1–27). Familiarity may dictate not only one's preference, as these studies show, but also judgments of validity, for instance, of statements (e.g., H. R. Arkes, C. Hackett, and L. Boehm, "The generality of the relation between familiarity and judged validity," *Journal of Behavioral Decision-Making, 2* (1989, 81–94). One would expect long-term potentiation in children as well (e.g., P. Aitken, D. Eadle, G. Hastings, and A. Haywood, "Predisposing effects of cigarette advertising on children's intention to smoke when older," *British Journal of Addiction, 86* (1991, 383–390), though appropriate long-term studies remain to be done (R. Kates, *Cigarette advertising and children: An experiment examining the effects of Joe Camel*, unpublished B.A. thesis, Princeton University, 1993).

The same potentiating effect of reencounter with the familiar might, however, provide an incidental argument in favor of cultural education for children. Although children may not appreciate all the things they learn about, they may later appreciate these things more than they would have otherwise, merely through having known about them earlier. Their interest may be piqued later toward going to see places, as well as toward appreciating them on site. There are limits, of course, to how much information children can or ought to deal with that is not of direct interest to them.

## Chapter Nine

1. S. Fraiberg, *The magic years: Understanding and handling the problems of early childhood* (New York: Scribner's, 1959), pp. 112–114.

2. C. N. Barnard, "Athens sojourn," *National Geographic Traveler Magazine* (1992, Nov./Dec.), p. 106.

3. W. Benjamin, "The work of art in the age of mechanical reproduction," in *Illuminations* (Trans., H. Zohn; New York: Schocken, 1968), p. 221. For related claims regarding forgery in art, see, for example, F. Sparshott, "The disappointed art lover" (in D. Dutton, Ed., *The forger's art*, Berkeley: University of California, 1983, 246–263) and other papers in the same volume, as well as M. Mothersill, *Beauty restored* (Oxford, England: Clarendon, 1984), p. 398. For relevant commentary on the offense created by the replacement of nature by artifice, see I. Kant, *The critique of judgment* (Trans., J. C. Meredith, Oxford, England: Clarendon, 1790/1986), pp. 158 and 162.

4. B. G. Harrison, "The master's hand," *The New York Times*, January 28, 1996, p. E13.

5. L. Alvarez, "Going, going, long gone," *The New York Times*, December 15, 1996, p. 51.

6. J. Perez, "Decay of a 20th century relic: What's the future of Auschwitz?" *The New York Times*, January 5, 1994, p. A6.

7. C. Trillin, "Anne of Red Hair: What do the Japanese see in Anne of Green Gables?" *The New Yorker*, August 5, 1996, p. 61.

8. D. Sobel, *Longitude: The true story of a lone genius who solved the greatest scientific problem of his time* (New York: Walker, 1995), p. 174.

9. G. Judson, "Two prisoners of history meet camera's captors: Frozen moments alter the lives of subjects of famous photos," *The New York Times*, October 11, 1995, pp. B1, B5.

## Chapter Ten

1. The following references provide pertinent observations, though not necessarily relevant interpretations, of children's spontaneous talk: K. Chukovsky, *From 2 to 4* (Berkeley: University of California Press, 1968); N. Isaacs, "Children's 'why' questions," in S. Isaacs, *Children's ways of knowing* (New York: Teachers College Press, 1974; original paper published in 1930); J. Piaget, *The language and thought of the child* (New York: Meridian, 1959; original work published in 1923), Chapter V; J. Sully, *Studies of childhood* (New York: Appleton, 1914), Chapters 2, 3, and 4. The evidence from modern databases, which is cited three paragraphs hence, is referenced in note 7 of this chapter.

2. J. Piaget, *The language and thought of the child* (1923; see also J. Piaget and B. Inhelder, *The child's conception of space*, New York: Norton, 1948) initiated interest in children's development of the ability both to know and to coordinate different perspectives. For samples of later work, see, for example, M. Donaldson, *Children's minds* (New York: Norton, 1979); J. H. Flavell, P. T. Botkin, C. L. Fry Jr., J. W. Wright, and P. Jarvis, *The development of role-taking and communication skills in children* (New York: Wiley, 1968).

3. In an early study of this behavior, children were shown a box of "Smarties" (Canadian M&Ms) and were asked to say what they saw. The box was hidden and then subsequently returned. Now, however, the box contained pencils rather than candy. The children were asked to say what was in the box now *and* to say what it contained before it was hidden.

When they were shown the box the first time, young children reported correctly that they saw Smarties. When the box was returned to them with pencils, some children said that they thought earlier that the box contained pencils. They reported their earlier belief incorrectly (A. Gopnik and J. Astington, "Children's understanding of representational changes and its relation to the understanding of false belief and the appearance-reality distinction," *Child Development*, 59, 1988, 26–37).

Piaget, *The moral judgment of the child* (New York: Free Press, 1932) and others (see references in Piaget, 1932) have referred to the related phenomenon of "pseudolying": children's nonmalicious alteration of the facts, to bring them into conformity with what others believe or expect of children. Sully (*Studies of childhood*, 1914) called this tendency an "illusion of memory" (p. 257). He reports the story of one child who, after running from a snake she had seen, told her brother and some of his friends that she had seen a "'sauger" (massasauga). The boys taunted back, "Well it didn't have a ring around its neck, did it?" Immediately the little girl saw "just such a serpent" in her mind (Sully, *Studies of childhood*, p. 257).

4. See, for example, selections in R. Fivush and J. A. Hudson (Eds.), *Knowing and remembering in young children* (New York: Cambridge University Press,

1990); C. A. Nelson (Ed.), *Memory and affect in development. The Minnesota Symposia on Child Psychology, Vol. 26* (Hillsdale, NJ: Erlbaum, 1993).

5. For example, Fraiberg, *The magic years: Understanding and handling the problems of early childhood* (New York: Scribner's, 1959) p. 114; K. Nelson, "Events, narratives, and memory: What develops?" in C. A. Nelson (Ed.), *Memory and affect in development. The Minnesota Symposia on Child Psychology, Vol. 26,* 1993).

6. For example, M. Donaldson, *Children's minds,* 1979; P. Mitchell and H. Lacohée, "Children's early understanding of false belief," *Cognition, 39* (1991, 107–127; M. Shatz and R. Gelman, "The development of communication skills: Modifications in the speech of young children as a function of the listener," *Monographs of the Society for Research in Child Development, 38* (No. 5), Serial No. 152, 1973.

7. A scan of the Childes Data Base ("The child language data exchange system," *Journal of Child Language, 12* (1985, 271–296), which contains several hundred hours of transcripts of the spontaneous speech of ten middle-class, English-speaking children, produced one occurrence of "exist," in a boy of 5 years, 9 months, who used the word incorrectly. (The children, drawn from different studies, were all observed longitudinally, from below 2 years of age to ages ranging from 2 1/2 to 8 years. Children normally reach minimal fluency in their native language by about 3 years. Four children in the database were observed until 5 years or more. One of these children was nearly 5, two were nearly 6, and one was nearly 8, at the termination of observation.) The *EDL Core Vocabularies,* which consist of lists of words drawn from extensive surveys of reading materials for children, recommend introduction of *exist* and *existence* at the sixth level, or approximately 11 years of age (*EDL core vocabularies in reading, mathematics, science and social studies,* ©Educational Development Labs, Inc., Columbia, SC, 1989). The EDL vocabulary lists are used for the design of graded reading materials and standardized tests, and also serve as general guides for teachers' classroom interaction.

The *EDL core vocabularies* list the root *real* for level 2, or children of approximately 7 years of age and above. This means that children of this age are expected to be able to read and understand the word. (See the next paragraph of text.)

8. Before her father could stop her, the child touched Princess Diana's skirt as she passed by at a reception, apparently, according to her father, because at that age she tended to touch people whom she liked. Princess Diana replied, "It's O.K. I'm real." (A. Leslie, personal communication, October 1995).

9. N. Link, personal communication, 1993.

10. Developmental psychological research demonstrates that as early as the fourth year, children can distinguish between what things are "really" and what they appear to be, and between what is real and what is in the mind; the boundary of the distinction changes with age. See, for instance, J. H. Flavell, E. R. Flavell, and F. L. Green, "Development of the appearance-reality distinction," *Cognitive Psychology, 15* (1985, 95–120; H. Wellman and D. Estes, "Early understanding of mental entities: A reexamination of childhood realism," *Child Development, 57* (1986, 910–923). See also pertinent observations in J. Piaget, *The child's conception of the world* (Totowa, NJ: Littlefield-Adams, 1960; original work published in

1926); J. Sully, *Studies of Childhood* (New York: Appleton, 1914), Chapters 3 and 4; Sugarman, *Piaget's construction of the child's reality* (New York: Cambridge University Press, 1987), Chapter 1.

11. Wordsworth, *Grosart, III*, p. 194, quoted in S. Gill, *William Wordsworth: A life* (New York: Oxford University Press, 1989), p. 33n.

12. S. Fraiberg, *The magic years . . .* (New York: Scribner's, 1959), p. 180.

13. G. B. Matthews, *Philosophy and the young child* (Cambridge, MA: Harvard University Press, 1980), p. 1.

14. E. A. Sherman, *Reminiscence and the self in old age* (New York: Springer, 1991).

15. W. Wordsworth, "Lines composed a few miles above Tintern Abbey, on revisiting the banks of the River Wye during a tour" (1798), lines 59–61, in T. Hutchinson and E. de Selincourt (Eds.), *Wordsworth: Poetical works* (New York: Oxford University Press, 1990), pp. 163–165).

16. W. Wordsworth, *The prelude* (1850), Book II, lines 315–318 (J. Wordsworth, M. N. Abrams, S. Gill, Eds., New York: Norton, 1979).

17. Rodger Brown, "Amusements: I'm in Yesterland!" in A. Heard (Ed.), "Sunday," *The New York Times*, May 25, 1997, sec. 6, p. 19.

## Chapter Eleven

1. He need not decide to ignore the negative evidence in either of these scenarios. He just does not consider it; see M. Johnston, "Self-deception and the nature of mind," in B. P. McLaughlin and A. O. Rorty (Eds.), *Perspectives on self-deception* (Berkeley: University of California Press, 1988), pp. 63–91) for related elaborations on the self-deceiver.

2. Philosophical problems may remain in the case of self-deception as well as in the case of ambivalence (H. Frankfurt, "The faintest passion," Presidential address delivered to the Eastern Division meeting of the American Philosophical Association, New York, December 29, 1991 (APA offprint). People may fall into spells of indecision under some circumstances in ways that violate some models of human rationality (e.g., D. A. Redelmeier and E. Shafir, "Medical decision making in situations that offer multiple alternatives," *Journal of the American Medical Association, 273* (1995, 302–305); A. Tversky and E. Shafir, "The disjunction effect in choice under uncertainty," *Psychological Science, 5* (1992, 305–309); see E. Shafir, "Intuitions about rationality and cognition," in K. I. Manktelow and D. E. Over (Eds.), *Rationality* (New York: Routledge, 1993), pp. 260–283, for additional discussion of the locus of people's reasoning in some of these choice situations.

3. S. Freud, *Introductory lectures on psychoanalysis* (1917), Chapter XVII, p. 270.

4. S. Freud, "The 'uncanny'" (1919), and S. Freud, *Jokes and their relation to the unconscious* (1905), Chapter 7, respectively.

5. S. Freud, *Jokes and their relation to the unconscious* (1905), Chapter 7.

6. D. Quinn, "An analysis of crying," unpublished manuscript, Princeton University, January 1997; D. Quinn, "Tears of joy: A developmental theory of the experience of the sublime," unpublished manuscript, Princeton University, April 1997.

7. G. Judson, "Two prisoners of history meet camera's captors: Frozen moments alter the lives of subjects of famous photos," *The New York Times*, October 11, 1995, pp. B1, B5.

8. N. Fein, "At the true 'Trial of the Century,'" *The New York Times*, October 13, 1996, sec. 5, p. 25.

# References

Aitken, P. (1976). *Television advertising and socialization to consumer roles* (Vol. 2). U.S. Department of Health and Human Services.

Aitken, P., Eadle, D., Hastings, G., & Haywood, A. (1991). Predisposing effects of cigarette advertising on children's intention to smoke when older. *British Journal of Addiction, 86,* 383–390.

Alvarez, L. (1996, December 15). Going, going, long gone. *The New York Times*, p. 51.

American Psychiatric Association. (1994). *Diagnostic and statistical manual of mental disorders* (4th ed.). Washington, DC: Author.

Aristotle. *The basic works of Aristotle* (R. McKeon, Ed.). New York: Random House, 1941.

Arkes, H. R., Hackett, C., & Boehm, L. (1989). The generality of the relation between familiarity and judged validity. *Journal of Behavioral Decision-Making, 2,* 81–94.

Ater, M. (1992). *The man Freud and monotheism.* Jerusalem: Magnes.

Austin, J. L. (1962). *Sense and sensibilia* (reconstructed from manuscript notes by G. J. Warnock). London: Oxford University Press.

Barnard, C. N. (1992, November/December). Athens sojourn. *National Geographic Traveler Magazine,* pp. 103–119.

Benjamin, W. (1968). The work of art in the age of mechanical reproduction. In *Illuminations* (Trans., H. Zohn). New York: Schocken Books, pp. 217–251.

Bower, T. G. R. (1974/1982). *Development in infancy* (2nd ed.). San Francisco: Freeman.

Brown, R. (1997, May 25). Amusements: I'm in Yesterland! In A. Heard (Ed.), "Sunday." *The New York Times,* sec. 6, p. 19.

Brown, R. (1973). *A first language: The early stages.* Cambridge, MA: Harvard University Press.

Bruner, J. S. (1983). *Child's talk: Learning to use language.* New York: Norton.

Bruner, J. S. (1990). *Acts of meaning.* Cambridge, MA: Harvard University Press.

Chalmers, D. (1996). *Conscious mind: In search of a fundamental theory.* New York: Oxford University Press.

Chuang-Tzŭ. (1974). *Inner chapters* (trans., G. Feng & J. English). New York: Vintage/Random House.

Chukovsky, K. (1968). *From 2 to 5.* Berkeley: University of California Press.

Cole, M., & Subbotsky, E. (1991, October 21–24). *The fate of stages past: Reflections on the heterogeneity of thinking from the perspective of cultural-historical psychology.* Paper presented at the symposium "The Cultural Environment in Psychology" honoring Ernst Boesch. Merlingen, Switzerland.

Cox, D., & Cox, A. (1988). What does familiarity breed? *Journal of Consumer Research, 15,* 111–116.

Darnton, J. (1994, March 20). Fiction is stranger than truth. *The New York Times*, sec. 4, p. 1.

Dennett, D. (1996). *Kinds of minds: Toward an understanding of consciousness.* New York: Basic Books.

Donaldson, M. (1979). *Children's minds.* New York: Norton.

Dutton, D. (Ed.). (1983). *The forger's art.* Berkeley: University of California Press.

*EDL core vocabularies in reading, mathematics, science and social studies.* (1989). Columbia, SC: Educational Development Labs.

Egan, T. (1991, July 3). 4 minutes of darkness puts a glow on Hawaii. *The New York Times*, sec. 1, p. 1.

Fein, N. (1996, October 13). At the true 'Trial of the Century.' *The New York Times*, sec. 5, p. 25.

Fivush, R., & Hudson, J. A. (Eds.). (1990). *Knowing and remembering in young children.* New York: Cambridge University Press.

Flannery, J. F. (1980). Freud's Acropolis revisited. *International Review of Psychoanalysis, 7,* 347–352.

Flavell, J. H., Botkin, P. T., Fry, C. L., Jr., Wright, J. W., & Jarvis, P. (1968). *The development of role-taking and communication skills in children.* New York: Wiley.

Flavell, J. H., Flavell, E. R., & Green, F. L. (1985). Development of the appearance-reality distinction, *Cognitive Psychology, 15,* 95–120.

Fraiberg, S. (1959). *The magic years: Understanding and handling the problems of early childhood.* New York: Scribner's.

Frankfurt, H. (1991, December 29). *The faintest passion.* Presidential address delivered to the Eastern Division meeting of the American Philosophical Association, New York (APA offprint).

Freemantle, M. (1996, January 29). Walking on air. *Chemical and Engineering News*, p. 34.

Freud, A. (1967). About losing and being lost. *The Psychoanalytic Study of the Child, 22,* 9–19.

Freud, S. (1900/1965). *The interpretation of dreams.* New York: Avon.

Freud, S. (1901/1989). *The psychopathology of everyday life.* New York: Norton.

Freud, S. (1905/1989). *Jokes and their relation to the unconscious.* New York: Norton.

Freud, S. (1908/1985). Creative writers and daydreaming. *Pelican Freud Library* (Vol. 14; A. Dickinson, Vol. Ed.). Harmonsworth, England: Penguin Books, pp. 129–141.

Freud, S. (1910/1961). *Five lectures on psychoanalysis.* New York: Norton.

Freud, S. (1911/1963). Formulations regarding two principles of mental functioning. In S. Freud, *General psychological theory* (P. Reiff, Ed.). New York: Collier-Macmillan, pp. 21–28.

Freud, S. (1913/1989). *Totem and taboo.* New York: Norton.

Freud, S. (1915a/1963). Instincts and their vicissitudes. In S. Freud, *General psychological theory* (P. Reiff, Ed.). New York: Collier/Macmillan, pp. 83–103.

Freud, S. (1915b/1963). The unconscious. In S. Freud, *General psychological theory* (P. Reiff, Ed.). New York: Collier/Macmillan, pp. 21–28.

Freud, S. (1916/1985). Some character-types met with in psychoanalytic work. *Pelican Freud Library* (Vol. 14; A. Dickson, Vol. Ed.). Harmonsworth, England: Penguin Books, pp. 291–320.

Freud, S. (1917/1977). *Introductory lectures on psychoanalysis.* New York: Liveright-Norton.

Freud, S. (1919/1985). The 'uncanny.' *Pelican Freud Library* (*Vol. 14*; A. Dickson, Vol. Ed.). Harmonsworth, England: Penguin Books, pp. 335–376.

Freud, S. (1920/1989). *Beyond the pleasure principle.* New York: Norton.

Freud, S. (1921/1989). *Group psychology and the analysis of the ego.* New York: Norton.

Freud, S. (1925/1963). Negation. In *General psychological theory* (P. Reiff, compiler). New York: Macmillan, pp. 213–217.

Freud, S. (1927/1961). *The future of an illusion.* New York: Norton.

Freud, S. (1936/1984). A disturbance of memory on the Acropolis: An open letter to Romain Rolland on the occasion of his seventieth birthday. (*Pelican Freud Library* [Vol. 11; A. Richards, Vol. Ed.]. Harmonsworth, England: Penguin Books, pp. 443–456, reprinted in this book).

[Freud, S. (1936/1950). Brief an Romain Rolland: Eine Erinnerungesstörung auf der Akropolis. *Gesammelte Werke, 16.* London: Imago, pp. 250–257.]

Freud, S. (1939/1949). *Moses and monotheism.* New York: Norton.

Freud, S. (1940/1989). *An outline of psychoanalysis.* New York: Norton.

Freud, S., & Breuer, J. (1893–1895/1983). *Studies on hysteria. Pelican Freud Library* (Vol. 3; A. Richards, Vol. Ed.). Harmonsworth, England: Penguin Books.

Fried, R. I. (1988), The Stendhal syndrome: Hyperkulturemia. *Ohio Medicine, 87,* 519–520.

Fromm-Reichmann, F. (1959). *Psychoanalysis and psychotherapy: Selected papers.* Chicago: University of Chicago Press.

Gardner, H. (1993). *Creating minds: An anatomy of creativity as seen through the lives of Freud, Einstein, Picasso, Stravinsky, Eliot, Graham, and Gandhi,* New York: Basic Books.

Gill, S. (1989). *William Wordsworth: A life.* New York: Oxford University Press.

Goldwyn, S. (Producer), McLeod, N. Z. (Director). (1947). The secret life of Walter Mitty. [Film]. (Available from Home Box Office)

Gopnik, A., & Astington, J. (1988). Children's understanding of representational changes and its relation to the understanding of false belief and the appearance-reality distinction. *Child Development, 59,* 26–37.

Gorman, J. (1997, April 29). Consciousness studies: From stream to flood. *The New York Times,* pp. C1, C5.

Gould, S. J. (1987). *An urchin in the storm.* New York: Norton.

Greenberg, J. R., & Mitchell, S. A. (1983). *Object relations in psychoanalytic theory.* Cambridge, MA: Harvard University Press.

The hand that shook the hand [photo inset]. (1996, March 15). *The New York Times,* p. B4.

Harris, P. L. (1983). Infant cognition. In P. H. Mussen (Series Ed.), *Handbook of child psychology: Vol. 2. Infancy and developmental psychobiology (4th ed.).* (M. M. Haith & J. J. Campos, Vol. Eds.), pp. 689–782.

Harrison, B. G. (1996, January 28). The master's hand. *The New York Times*, p. E13.

Harrison, I. B. (1966). A reconsideration of Freud's "A disturbance of memory on the Acropolis" in relation to identity disturbance. *Journal of the American Psychoanalytic Association, 14,* 518–527.

Horner, M. (1970). Femininity and successful achievement: A basic inconsistency. In J. Bardwick, E. L. Douvan, M. S. Horner, & D. Gutmann (Eds.), *Feminine personality and conflict.* Pacific Grove, CA: Brooks/Cole, pp. 45–74.

Howe, N. (1993). Reading places. *Yale Review, 81,* 60–73.

Isaacs, N. (1930/1974). Children's "why" questions. In N. Isaacs, *Children's ways of knowing.* New York: Teachers College Press, pp. 13–64.

Johnson, M. K., & Raye, C. L. (1981). Reality monitoring. *Psychological Review, 88,* 67–85.

Johnston, M. (1988). Self-deception and the nature of mind. In B. P. McLaughlin & A. O. Rorty (Eds.), *Perspectives on self-deception.* Berkeley: University of California Press, pp. 63–91.

Judson, G. (1995, October 11). Two prisoners of history meet camera's captors: Frozen moments alter the lives of subjects of famous photos. *The New York Times*, pp. B1, B5.

Kant, I. (1790/1986). *The critique of judgment* (trans., J. C. Meredith). Oxford, England: Clarendon.

Kates, R. (1993). *Cigarette advertising and children: An experiment examining the effects of Joe Camel.* Unpublished B.A. thesis, Princeton University, Princeton, NJ.

Keillor, G. (1992). [Recorded by G. Keillor and F. von Stade]. *Songs of the cat* [cassette]. New York: BMG Classics.

Kleinfield, N. R. (1995, May 1). Death scene now is lure for curious. *The New York Times*, pp. A1 and B7.

Laing, R. D. (1965). *The divided self.* Baltimore: Penguin.

Lively, P. (1989). *Moon tiger.* New York: HarperCollins.

Loftus, E. F. (1979). The malleability of human memory. *American Scientist, 67,* 312–320.

MacWhinney, B., & Snow, C. (1985). The child language data exchange system. *Journal of Child Language, 12,* 271–296.

Masson, J. M., & Masson, T. C. (1978). Buried memories on the Acropolis: Freud's response to mysticism and anti-Semitism. *International Journal of Psychoanalysis, 59,* 199–208.

Matthews, G. B. (1980). *Philosophy and the young child.* Cambridge, MA: Harvard University Press.

Mitchell, P., & Lacohée, H. (1991). Children's early understanding of false belief. *Cognition, 39,* 107–127.

Mothersill, M. (1984). *Beauty restored.* Oxford, England: Clarendon.

Nelson, C. A. (Ed.). (1993). *Memory and affect in development. The Minnesota Symposia on Child Psychology* (Vol. 26). Hillsdale, NJ: Erlbaum.

Nelson, K. (1985). *Making sense: The acquisition of shared meaning.* Orlando, FL: Academic Press.

Nelson, K. (1993). Events, narratives, and memory: What develops? In C. A. Nelson (Ed.), *Memory and affect in development. The Minnesota Symposia on Child Psychology* (Vol. 26). Hillsdale, NJ: Erlbaum, pp. 1–24.

Norem, J., & Cantor, N. (1986). Defensive pessimism: Harnessing anxiety as motivation. *Journal of Personality and Social Psychology, 51,* 1208–1217.

Penrose, R. (1989). *The emperor's new mind.* New York: Oxford University Press.

Perez, J. (1994, January 5). Decay of a 20th century relic: What's the future of Auschwitz? *The New York Times,* p. A6.

Piaget, J. (1923/1959). *The language and thought of the child.* New York: Meridian.

Piaget, J. (1926/1960). *The child's conception of the world.* Totowa, NJ: Littlefield-Adams.

Piaget, J. (1932/1965). *The moral judgment of the child.* New York: Free Press.

Piaget J. (1936/1952). *The origins of intelligence in children.* New York: International Universities Press.

Piaget, J. (1937/1954). *The construction of reality in the child.* New York: Basic Books.

Piaget, J., & Inhelder, B. (1948/1956). *The child's conception of space.* New York: Norton.

Pomorsky, K. (Ed.). (1987). Language, poetry, and poetics: The generation of the 1890s: Jakobson, Trubetskoy, Majakovsky. *Proceedings of the first Roman Jakobson Colloquium at the Massachusetts Institute of Technology, October 5–6, 1984.* Berlin & New York: de Gruyter.

Quinn, D. (1997a). *An analysis of crying.* Unpublished manuscript, Princeton University, Princeton, NJ.

Quinn, D. (1997b). *Tears of joy: A developmental theory of the experience of the sublime.* Unpublished manuscript, Princeton University, Princeton, NJ.

Redelmeier, D. A., & Shafir, E. (1995). Medication decision making in situations that offer multiple alternatives. *Journal of the American Medical Association, 273,* 302–305.

Riccio, D. C., Rabinowitz, V. C., & Axelrod, S. (1994). Memory: When less is more. *American Psychologist, 49,* 917–926.

Russell, B. (1988). Knowledge by acquisition and knowledge by description. In *The problems of philosophy.* Buffalo, NY: Prometheus, pp. 46–59

Schorske, C. (1993, May 27). Freud's Egyptian dig. *The New York Review of Books, XLI*(10), 35–40.

Searle, J. R. (1975). Indirect speech acts. In P. Cole & J. L. Morgan (Eds.), *Syntax and semantics: Vol. 3. Speech acts.* New York: Academic Press, pp. 59–82.

Searle, J. R. (1992). *The rediscovery of the mind.* Cambridge, MA: MIT Press.

Shafir, E. (1993). Intuitions about rationality and cognition. In K. I. Manktelow & D. E. Over (Eds.), *Rationality.* New York: Routledge, pp. 260–283.

Shatz, M., & Gelman, R. (1973). The development of communication skills: Modifications in the speech of young children as a function of the listener. *Monographs of the Society for Research in Child Development, 38*(5, Serial No. 152).

Sherman, E. A. (1991). *Reminiscence and the self in old age.* New York: Springer.

Slochower, H. (1970). Freud's *déjà vu* on the Acropolis: A symbolic relic of '*mater nuda.*' *Psychoanalytic Quarterly, 39,* 90–102.

Sno, H. M., & Linszen, D. H. (1990). The *déjà vu* experience: Remembrance of things past? *American Journal of Psychiatry, 147,* 1587–1595.

Sobel, D. (1995). *Longitude: The true story of a lone genius who solved the greatest scientific problem of his time.* New York: Walker.

Sparshott, F. (1983). The disappointed art lover. In D. Dutton (Ed.), *The forger's art.* Berkeley: University of California Press, pp. 246–263.

Sugarman, S. (1981). The cognitive basis of classification in very young children: An analysis of object ordering trends. *Child Development, 52,* 1172–1178.

Sugarman, S. (1983). *Children's early thought: Developments in classification.* New York: Cambridge University Press.

Sugarman, S. (1987). *Piaget's construction of the child's reality.* New York: Cambridge University Press.

Sugarman, S. (1993). Piaget on the origins of mind: A problem in accounting for the development of mental capacities. In E. Dromi (Ed.), *Language and cognition: A developmental perspective.* Norwood, NJ: Ablex.

Sully, J. (1914). *Studies of childhood.* New York: Appleton.

Swann, W. (1996). *Self traps: The elusive quest for higher self-esteem.* New York: Freeman.

Trillin, C. (1996. August 5). Anne of Red Hair: What do the Japanese see in Anne of Green Gables? *The New Yorker,* pp. 56–61.

Tversky, A., & Shafir, E. (1992). The disjunction effect in choice under uncertainty. *Psychological Science, 5,* 305–309.

Watson, J. S. (1973). Smiling, cooing, and the game. *Merrill-Palmer Quarterly, 18,* 323–339.

Wellman, H., & Estes, D. (1986). Early understanding of mental entities: A reexamination of childhood realism. *Child Development, 57,* 910–923.

Wittgenstein, L. (1958). *Philosophical investigations* (3rd ed.). New York: Macmillan.

Wordsworth, W. (1798/1990). Lines composed a few miles above Tintern Abbey, on revisiting the banks of the River Wye during a tour. In T. Hutchinson & E. de Selincourt (Eds.), *Wordsworth: Poetical works.* New York: Oxford University Press, pp. 163–165.

Wordsworth, W. (1807/1990). Ode: Intimations of immortality from recollections of early childhood. In T. Hutchinson & E. de Selincourt (Eds.), *Wordsworth: Poetical works.* New York: Oxford University Press, pp. 460–462.

Wordsworth, W. (1850/1979). *The Prelude* (J. Wordsworth, M. H. Abrams, S. Gill, Eds.). New York: Norton.

Zajonc, R. (1968). Attitudinal effects of mere exposure. *Journal of Personality and Social Psychology, Monograph Supplement, 9,* No. 2, Part 2, 1–27.

# Index

Accident
  as basis for Freud's thought, 10
  of behavior, 29
Adult. *See also* Self
  contexts for return to childlike in,
    50–55, 67–68
  experience of reencounter by, 48–50
Advertising, 54
Affirmation. *See also* Confirmation
  context requirements for, 47
  Freud's thought as, 24, 70
  irrational doubt dispelled with,
    37–38
Airplane travel, childlike experience of,
  48
Airport scenario, 36–38
Alarm, irrational doubt causing, 36–37,
  46
Ambiguity
  in childlike experience, 53
  in presentation of historical reality,
    22
  in verbal expression, 10
Ambivalence, 74, 92(n2)
Ancients, experience of reality by, 11
*Anne of Green Gables* (Montgomery),
  8, 59, 61, 77
Aristotle, 54
Art, imitative aspect of, 54
Atlantic Ocean, 41, 48, 91(n7)
"Aura" (after W. Benjamin), 35, 91(n1,
  Ch. 6)
Auschwitz, 58–59, 61
Austin, J. L., 24
Authenticity, effect of on object's
  appreciation, 58–59
Awe. *See also* Surprise
  context of affecting experience,
    19–20, 25

in experience of the real thing, 8, 58
feelings of unreality in experience of,
  35–36
in mundane environment, 41–42, 46

Behaviors
  conflict as basis for, 74–75
  context for, 29
Beliefs
  of children, 66
  "good grounds" for, 27
  seeing required for, 57, 67
Benjamin, Walter, 58
Bird watching, 53, 65
Book review, 42–43
Bryan, William Jennings, 9
Buchanan, Pat, 25

Certainty. *See also* Doubt;
  Uncertainty
  as component of Freud's thought, 6,
    45
Child mentality, reassertion of in adult
  experience, 22
Childes Data Base, 96(n7)
Childhood
  central issues of, 53
  uncertainty about reality in, 9, 13,
    22, 47, 57
Childlike
  adult contexts for return to, 50–55
  quality of in Freud's thought, 47–55,
    55–57, 59, 64, 65–68, 71
  repression of, 75
Children
  cognitive-developmental occupations
    of, 52
  desire of for literal experience,
    49–50, 59

experience of reencounter by, 49–50, 64

and Freud's thought, 65–68, 73

Chuang-Tzŭ, 89 (n4, Ch. 3)

Circumlocution, compared to Freud's thought, 26

Cliché, compared to Freud's thought, 27

Closure, experience in Freud's thought, 70

Cognitive-developmental psychology, 52

Comic sense, 75

Common-sense meaning
  of Freud's thought, 17–24, 25, 45
  summary of, 27–28

Confirmation. *See also* Affirmation
  as basis of Freud's thought, 76
  coincidence of with wonder, 38
  of experience, 68–72, 73
  for previously doubted reality, 5

Conflict
  as basis of behaviors, 74–75
  not a factor in Freud's thought, 73–74

Context
  antecedent doubt in, 22–23
  for Freud's thought, 9–10, 23, 30, 47
  literal interpretations for, 23–24
  no doubt about reality in, 19–20
  open questioning of reality in, 21–22

Crying. *See also* Feelings
  significance of experience triggering, 60–61, 76–77

Cultural education, 94(n29)

Cynical expressions, compared to Freud's thought, 18

Dachstein, 49–50, 59

Darrow, Clarence, 9

Dayton (TN), 9

Defense
  against guilt, 7, 29, 30–31, 35, 38, 83, 90(n6)
  derealization function for, 14
  recurrence of, 32–33
  return to childlike as, 52

Devensive pessimism, 90–91(n13)

Déjà vu, compared to Freud's thought, 7–8, 83

Denial. *See also* Repudiation
  and displacement, 32–33, 75
  relationship to guilt, 6, 31
  of truth, 32

Depersonalization disorder, compared to Freud's thought, 7–8, 14, 82–83

Depression. *See also* Feelings
  as precursor to Freud's thought, 13, 14, 30

"Depth of processing" experience, 21

Derealization. *See also* Unreality
  compared to Freud's thought, 7
  experience of as basis for Freud's thought, 14–15, 29, 30, 35–36, 82–83, 90(n6)

Diana, Princess of Wales, 67, 96(n8)

Direct contact. *See also* Experience of reality
  influence of on Freud's thought, 44–45

Displacement
  defined, 90(n4)
  and denial, 32–33, 75
  of doubt, 14, 29, 30
  of incredulity, 14

"Disturbance of Memory on the Acropolis, A," 5–6, 79–86

Doubt. *See also* Uncertainty
  antecedent, 22–23
  childhood experience of, 9, 13
  as factor in Freud's thought, 27, 35, 45, 68–69, 80, 82, 84
  hypothetical, 20
  irrational, 35–37
  of mundane reality, 44
  not a factor in Freud's thought, 19–20
  open question allowed for, 21–22
  origin of in past, 14
  reassurance for, 27
  residual, 53
  surprise at discovery of, 5

Dreams, 29, 67, 75

Emotions. *See* Feelings

Evidence of reality. *See also* Experience of reality; Reality
affect on children, 66
affected by previous information, 54–55
compared to second-hand information, 13
surprise at discovery of, 6

"Exist/Existence," as vocabulary items, 66, 96(n7)

Expectations, congruence of reality with, 37

Experience
compared to second-hand information, 13, 20, 22
confirmation of, 68–72
doubt of affecting Freud's thought, 20
validation of, 71

Experience of reality. *See also* Evidence of reality; Reality
affected by previous information, 54–55
in contact with the real thing, 57–58
"depth of processing" affecting, 21
influence of on Freud's thought, 44–45
literal basis for, 18–19, 71
miraculous overtone for, 39–41
perceptual differences affecting, 17–18
provocation of with symbol, 44

Face-value meaning. *See* Common sense meaning

Fakes. *See also* Fiction
disillusion in experience of, 58
replicas, 53
transformation of into reality, 8

False memory. *See also* Memory
compared to Freud's thought, 7–8
considering of by skeptic, 20

Familiar
appreciation of, 54–55, 76
experienced as strange, 7

personal experience of, 59–61
reflection on, 67
role of in artistic appreciation, 54

Fear, affect on wish fulfillment, 14

Feelings. *See also* Crying; Depression
evocation of through image, 60–64
expressed in contact with the real thing, 59, 60
in initial experience of reality, 9, 77
relation of to Freud's thought, 9–10, 18, 38

Fein, Naomi, 8–9, 77

Fiction. *See also* Fakes
experienced as reality, 8, 59

Forgery, 94(n3)
in art, 58–59
*See also* Authenticity; Replicas; Fakes

Fossils, 58

Frame of reference, shifts in affecting recognition, 70

French Embassy, 8, 58

Freud, S.
on the Acropolis, 5, 8, 13–15, 21, 22, 36, 37, 64, 79–84, 88–89(n22), 89(n5), 90(n6)
analysis of "Freud's thought," 6, 7, 9, 13–15, 17–18, 29–33, 35, 74, 79–86, 88–89(n22)
conflict, role of in theory, 74–75
Dachstein, sojourn to, 49–50, 59, 66
delusional jealousy, analysis of, 30–31, 74–75
on displacement, 90(n4)
on fiction, 92(n13)
on jokes, 92(n13)
on method of inquiry, 1, 3, 17–18, 74–76
on object relations, 52–53, 75–76
on primary process thought, 90(n4)
repression, role of in theory, 75–76
on the return of the childlike, 52, 75–76

Frustration, 14

Games, rediscovery of pleasure in, 52–53

Giuliani, Rudolph, 50–51

Grand Canyon, 9, 22–23, 58, 61

Guilt
  affect on wish fulfillment, 14
  at surpassing one's parents, 6, 7, 15,
    30
  as basis for denial of experience, 6, 7,
    14–15, 29, 30–31, 32–33, 36
  relief for with delusional jealousy,
    30–31

Handshake, 50–51. *See also* Touch
Harrison, Barbara Grizzuti, 58
Harrison, John, 60
Heaney, Seamus, 32
Hedge example, 42, 44, 64, 69
Hitler, Adolph, 25
Howe, Nicholas, 54
Hyperbole, compared to Freud's
    thought, 25–26

Idiosyncracy, 9, 10
Illness, wish fulfillment affecting, 13–14
Illusion
  compared to reality, 70–71
  reality experienced as, 7
Images. *See also* Photographs
  confirmation of reality in viewing of,
    50
  as force in self-development,
    60–61
  pleasure experienced in encounter
    with, 53–54
  veracity or falsehood of, compared,
    58
Incredulity
  displacement of, 14
  mystique of objects affecting, 18
  in repudiation of reality, 13–14
Indecision, 74
Indirect discourse, compared to Freud's
    thought, 26
Infants. *See* children

Japanese tourists, 8, 59, 77
Jealousy, delusional, 30–31, 74–75
Jokes, 2, 3, 54, 73, 92(n13), 93(n27)
Judgment, formation of not typical of
    child, 66

Kant, I., 94(n3)
Kent State University, 60–63
Kim Phuc, 60–63
Knowledge
  affecting appreciation of known
    object, 55, 58
  fallibility of, 20

Language use, compared to Freud's
    thought, 26–27
Laughter, 75
Legends, 20
Legette, Kyronne, 50–51
Linguistic tropes, compared to Freud's
    thought, 25–26
Literal expression, Freud's thought as,
    18–23
Lively, Penelope, 6, 18
Loch Ness monster, 5, 20, 39, 80
*Longitude* (Sobel), 60
Loose usage, compared to Freud's
    thought, 26–27
Loss
  affect of on reencounter, 48–49, 53, 70
  infant's experience of, 52–53
Love object. *See also* Object
  rediscovery of, 52–53
Luxembourg, 22–23

Map, relation of to literal reality, 43,
    45, 50, 71
Marble Mountains, 43, 64
Meaning
  affirmation of past enhancing, 71
  discovery of, 10
  face-value, 17
Memory. *See also* False memory
  affecting Freud's thought, 64, 68
  as basis for déjà vu, 7
  of illusion, 71–72
  repression of, 14
Mental processes, relation of to Freud's
    thought, 9–10
Mental "tricks," 6–7
Metaphor
  compared to Freud's thought, 25
  for miraculous experience, 40

Michelangelo's David, 8, 58, 61
*Mimesis,* 54
Miracles, responses to, compared to
    mundane reality, 39–41
Mona Lisa, 58, 59
Montgomery, Lucy Maude, 8, 59
*Moon Tiger,* 6
Morality, 2
Mummy, 40
Mundane. *See also* "Uninteresting
    commonplace"
    compared to miraculous, 41–46
    perception of as miraculous,
        40–41
    presentation of in Freud's thought,
        23–24, 39, 47
    transformation of into significance,
        8, 64

Narcissism, benign, 54
Neurotic symptoms, 29, 53, 68
    conflict-repression model for, 75
    influence on Freud's thought, 35–38
Newspaper story
    as confirmation of reality, 50, 54
    as precursor to actual experience, 6,
        54
*The New York Times Book Review,*
    42–43, 44, 45, 49
Nobel Prize, 32
Nonliteral assertions, compared to
    Freud's thought, 27–28
Nostalgia, 69, 72, 76

Object. *See also* Reality; The real thing
    childlike experience of, 48, 66–67
    establishment of "permanence" of,
        52
    loss of, experienced by infant,
        52–53
    love object, 52–53
    matching of words to, 50
    open questioning of, 21–22
    Freud's thought's requirements for,
        44, 45–46, 68–69
Obvious. *See* Mundane
Oklahoma City bombing, 8

Paradox, force of in Freud's thought, 5,
    11, 71–77
Past
    affect of in Freud's thought, 61–64,
        68–69
    affect of uncertainty in, 14
    awe experienced in, 19
    children's beliefs regarding, 66
    confirmation of experience in,
        68–72
    displacement of doubt into, 14, 29,
        30
    experienced in present, 55, 76
    personal connection with, 59–61, 69
    re-evocation of, 58–59
Pathological disturbances
    compared to Freud's thought, 7–8
    conflict-repression model for, 75
    return to childlike as, 52
Perceptual differences, role of in
    experience of Freud's thought,
        17–18
Permanence, childlike perception of,
    52
Photographs. *See also* Images
    confirmation of reality in viewing of,
        6, 43, 64
    as force in self-development, 60–61
    as stimulus to experience reality, 8
Piaget, Jean, 52
Pleasure
    in experience of mundane, 42–43
    reduction of by irrational doubt,
        36–37
    in reencounter, 49–50, 52–53, 55
    release of doubt obtained through,
        38
Present, resolution of past with, 72
Primary process thought, 90(n4)
Prince Edward Island, 8, 59, 77
Projection, of repressed material, 31
Psychodynamic analysis, of Freud's
    thought, 29–33, 74
Psychological context. *See* Context

"Real," as a vocabulary item, 96(n7)
Real/really, force of in context, 24

Reality. *See also* Evidence of reality;
     Experience of reality
  alternative representations of, 44–45
  assertion of in Freud's thought, 8
  childlike experience of, 48, 66–67
  confirmation of with photographs, 6,
     43
  congruence with expectations, 37
  in daily life, 8
  derealization experience of, 7
  doubt about existence of, 5
  experience of in ancient times, 11
  fascination with, 9
  historical, 60–63
  of image, compared to falsehood, 58
  incredulous repudiation of, 13–14
  loss and rediscovery of, 52–53
  not contemplated by children, 66–67
  open questioning of, 21–22
  personal contact with, 8–9
  reencounter with, 48–49
Real thing, the. *See also* Object
  significance of contact with, 57–61
Recognition
  frame of reference shifts affecting, 70
  pleasure experienced in, 53, 54
  "shock of," 54, 60
Recurrence, of Freud's thought, 10,
     32–33
Reencounter
  childlike interest in, 48, 52, 53, 55,
     57, 64, 76
  with historical reality, 61
  unreality experienced in, 68–70
Reflection, as characteristic of Freud's
     thought, 66, 67
Replicas, 53. *See also* Authenticity;
     Fakes
Repression
  conflict-repression model for,
     75–76
  relation to defense mechanisms, 31
Repudiation. *See also* Denial
  of reality, 13–15
  of success, 15
Rolland, Romain, 5–6
Russell, Bertrand, 20

Santa Claus, 67
Saturn, 40–41
School, presentation of reality in, 5, 26,
     42, 61, 64, 69
Scopes, John, 9
Scopes "monkey trial," 9, 77
Sea of Galilee, 42, 43, 44
"The Secret Life of Walter Mitty"
     (Thurber), 40
Self
  connection of to reality, 72, 76
  depersonalization disorder
     experience of, 7, 14
  integration of, 76–77
  re-creation of, 59, 69
Self-abhorrence, punishment for, 31,
     74–75
Self-awareness, lack of in performance
     of behavior, 29
Self-deception, conflict affecting, 74,
     97(n2)
Self-importance
  continuous desire for, 55
  feeling of in Freud's thought, 50
Sense impressions, 9–10
Skepticism. *See also* Doubt
  about experience, 20, 24
"Slap of responsibility," 60, 77
Slips of the tongue, 29
Sobel, Dava, 60, 77
Solar eclipses, 40
South Street Seaport, 48
Sphinx, 6, 18
Stendhal Syndrome, the, 90(n6)
Success
  destructive force of, 14, 81
  displacement of affirmation of, 32
  essence of, 15, 84
  repudiation of, 15
Surprise. *See also* Awe; Wonder
  in childlike enthusiasm, 52, 73
  in discovery of known reality, 5–6,
     11, 65–66, 68–69, 73
  experience of past as, 70
Symbol
  children's fascination with, 57
  provocation of reality with, 44

Thought, Freud's
  as childlike. *See* Childlike, quality of
    in Freud's thought
  and children, 65–68
  compared with other mental
    "tricks," 7
  as a confirmation of experience,
    68–72
  demographics for, 11
  and derealization. *See* Derealization
  description of, 6
  face-value meaning of, 17–24, 27–28
  and fascination with the real, 8–9,
    57–64
  as figurative expression, 25–28
  Freud's analysis of. *See* Freud,
    analysis of "Freud's thought"
  general case analysis of, 32–33
  and history of the object, 58–59
  as literal expression, 18–23
  as mundane, 39–41
  and nostalgia. *See* Nostalgia
  as paradoxical. *See* Paradox, force of
    in Freud's thought
  and the personal past, 59–61
  popular occurrence of, 11
  psychodynamic account of. *See*
    Freud, analysis of "Freud's
    thought"
  psychological context for, 9–10
  reencounter, basis in. *See*
    Reencounter
Thurber, James, 40
Time passage, effect of on reencounter,
  48–49, 53, 68–70
Tomb of the Unknown Soldier, 55
"Too good to be true," experience of
  affecting Freud's thought, 13,
  32–33, 76–77, 81
Touch
  compared to verbal/visual
    description, 17
  in handshake, 50–51
  relation to authenticity, 58–59, 67
Town meeting example, 42, 44, 45

Travel literature, 54
Tropes, discourse and linguistic, 25–26

Uncanny, feeling of the, 2
Uncertainty. *See also* Certainty;
  Doubt
  as basis for Freud's thought, 9, 11,
    68, 75–76
  lack of in Freud's thought, 26–27
"Uninteresting commonplace". *See also*
  Mundane
  rejection of as basis for Freud's
    thought, 13
Unreality. *See also* Derealization
  displacement of feelings about,
    30–31
  experience of with real objects,
    22–23, 69–70

Vecchio, Mary Ann, 60–63
Verbal expression, ambiguity of
  meaning in, 10
Vietnam War, 60–63

Weiss, Werner, 70
What if . . . scenario, 36–38
Wildflowers, 53
Wish fulfillment, affecting illness,
  13–14
Wishful thinking, as basis for false
  memory, 7
Wittgenstein, Ludwig, 88(n20)
Wonder. *See also* Surprise
  childlike experience of, 52
  coincidence of with confirmation, 38
  examples of objects for, 40–41
  factors affecting, 19–20, 45
  in indirect discourse, 26
  relationship of to Freud's thought,
    22–23
Words, children's fascination with, 57
Wordsworth, William, 69, 93(n14)

Yam Kinneret, 42, 43
"Yesterland" (Weiss), 70

Printed and bound by CPI Group (UK) Ltd, Croydon, CR0 4YY

23/10/2024

01778240-0014